U0211770

原版影印说明

1.《凝聚态物质与材料数据手册》（6册）是*Springer Handbook of Condensed Matter and Materials Data* 的影印版。为使用方便，由原版1卷改为6册：

第1册 通用表和元素

第2册 材料类：金属材料

第3册 材料类：非金属材料

第4册 功能材料：半导体和超导体

第5册 功能材料：磁性材料、电介质、铁电体和反铁电体

第6册 特种结构

2. 全书目录、作者信息、缩略语表、索引在各册均完整呈现。

本手册数据全面准确，1 025个图和914个表使查阅更加方便，是非常实用的案头参考书，适于材料及相关专业本科生、研究生、专业研究人员使用。

材料科学与工程图书工作室

联系电话 0451-86412421

0451-86414559

邮　　箱 yh_bj@aliyun.com

xuyaying81823@gmail.com

zhxh6414559@aliyun.com

Springer 手册精选原版系列

凝聚态物质与材料数据手册

通用表和元素

【第1册】

Springer
Handbook of
Condensed Matter and Materials Data

W.Martienssen

H.Warlimont

Editors

黑版贸审字08-2014-009号

图书在版编目（CIP）数据

凝聚态物质与材料数据手册．第1册，通用表和元素：英文/（德）马蒂安森（Martienssen，W.），（德）沃利蒙特（Warlimont，H.）主编．—哈尔滨：哈尔滨工业大学出版社，2014.3
（Springer手册精选原版系列）
ISBN 978-7-5603-4455-3

Ⅰ．①凝…　Ⅱ．①马…　②沃…　Ⅲ．①凝聚态-材料-技术手册-英文　Ⅳ．①O469-62　②TB3-62

中国版本图书馆CIP数据核字（2013）第291541号

材料科学与工程
图书工作室

责任编辑　杨　桦　许雅莹　张秀华
出版发行　哈尔滨工业大学出版社
社　　址　哈尔滨市南岗区复华四道街10号　邮编 150006
传　　真　0451-86414749
网　　址　http://hitpress.hit.edu.cn
印　　刷　哈尔滨市石桥印务有限公司
开　　本　787mm × 960mm　1/16　印张 13.25
版　　次　2014年3月第1版　2014年3月第1次印刷
书　　号　ISBN 978-7-5603-4455-3
定　　价　58.00元

Springer Handbook of Condensed Matter and Materials Data

W. Martienssen and H. Warlimont (Eds.)

With 1025 Figures and 914 Tables

Springer Handbook provides a concise compilation of approved key information on methods of research, general principles, and functional relationships in physics and engineering. The world's leading experts in the fields of physics and engineering will be assigned by one or several renowned editors to write the chapters comprising each volume. The content is selected by these experts from Springer sources (books, journals, online content) and other systematic and approved recent publications of physical and technical information.

The volumes will be designed to be useful as readable desk reference book to give a fast and comprehensive overview and easy retrieval of essential reliable key information, including tables, graphs, and bibliographies. References to extensive sources are provided.

Preface

The Springer Handbook of Condensed Matter and Materials Data is the realization of a new concept in reference literature, which combines introductory and explanatory texts with a compilation of selected data and functional relationships from the fields of solid-state physics and materials in a single volume. The data have been extracted from various specialized and more comprehensive data sources, in particular the Landolt–Börnstein data collection, as well as more recent publications. This Handbook is designed to be used as a desktop reference book for fast and easy finding of essential information and reliable key data. References to more extensive data sources are provided in each section. The main users of this new Handbook are envisaged to be students, scientists, engineers, and other knowledge-seeking persons interested and engaged in the fields of solid-state sciences and materials technologies.

The editors have striven to find authors for the individual sections who were experienced in the full breadth of their subject field and ready to provide succinct accounts in the form of both descriptive text and representative data. It goes without saying that the sections represent the individual approaches of the authors to their subject and their understanding of this task. Accordingly, the sections vary somewhat in character. While some editorial influence was exercised, the flexibility that we have shown is deliberate. The editors are grateful to all of the authors for their readiness to provide a contribution, and to cooperate in delivering their manuscripts and by accepting essentially all alterations which the editors requested to achieve a reasonably coherent presentation.

An onerous task such as this could not have been completed without encouragement and support from the publisher. Springer has entrusted us with this novel project, and Dr. Hubertus von Riedesel has been a persistent but patient reminder and promoter of our work throughout. Dr. Rainer Poerschke has accompanied and helped the editors constantly with his professional attitude and very personable style during the process of developing the concept, soliciting authors, and dealing with technical matters. In the later stages, Dr. Werner Skolaut became a relentless and hard-working member of our team with his painstaking contribution to technically editing the authors' manuscripts and linking the editors' work with the copy editing and production of the book.

We should also like to thank our families for having graciously tolerated the many hours we have spent in working on this publication.

Prof. Werner Martienssen

Prof. Hans Warlimont

We hope that the users of this Handbook, whose needs we have tried to anticipate, will find it helpful and informative. In view of the novelty of the approach and any possible inadvertent deficiencies which this first edition may contain, we shall be grateful for any criticisms and suggestions which could help to improve subsequent editions so that they will serve the expectations of the users even better and more completely.

September 2004
Frankfurt am Main, Dresden

Werner Martienssen,
Hans Warlimont

About the Authors

Wolf Assmus

Johann Wolfgang Goethe-University
Physics Department
Frankfurt am Main, Germany
assmus@physik.uni-frankfurt.de
http://www.rz.uni-frankfurt.de/piweb/kmlab/Leiter.html

Chapter 1.3

Dr. Wolf Assmus (Kucera Professor) is Professor of Physics at the University of Frankfurt and Dean of the Physics-Faculty. He is a solid state physicist, especially interested in materials research and crystal growth. His main research fields are: materials with high electronic correlation, quasicrystals, materials with extremely high melting temperatures, magnetism, and superconductivity.

Stefan Brühne

Johann Wolfgang Goethe-University
Physics Department
Frankfurt am Main, Germany
bruehne@physik.uni-frankfurt.de

Chapter 1.3

Dr. Stefan Brühne, née Mahne, a chemist by education in Germany and England, received his PhD in 1994 from Dortmund University, Germany, on giant cell crystal structures in the Al–Ta system. Following a post doc position at the Materials Department (Crystallography) at ETH Zurich he spent seven years in the ceramics industry. His main activity was R&D of glasses, frits and pigments for high-temperature applications, thereby establishing design of experiment (DoE) techniques. Since 2002, at the Institute of Physics at Frankfurt University he has been investigated X-ray structure determination of quasicrystalline, highly complex and disordered intermetallic materials.

Fabrice Charra

Commissariat à l'Énergie Atomique, Saclay
Département de Recherche sur l'État Condensé, les Atomes et les Molécules
Gif-sur-Yvette, France
fabrice.charra@cea.fr
http://www-drecam.cea.fr/spcsi/

Chapter 5.3

Fabrice Charra conducts researche in the emerging field of nanophotonics, in the surface physics laboratory of CEA/Saclay. The emphasis of his work is on light emission and absorption form single nanoscale molecular systems. His area of expertise also extends to nonlinear optics, a domain to which he contributed several advances in the applications of organic materials.

Gianfranco Chiarotti

University of Rome "Tor Vergata"
Department of Physics
Roma, Italy
chiarotti@roma2.infn.it

Chapter 5.2

Gianfranco Chiarotti is Professor Emeritus, formerly Professor of General Physics, Fellow of the American Physical Society, fellow of the Italian National Academy (Accademia Nazionale dei Lincei). He was Chairman of the Physics Committee of the National Research Council (1988–1994), Chair Franqui at the University of Liège (1975), Assistant Professor at the University of Illinois (1955–1957), Editor of the journal Physics of Solid Surfaces, and Landolt–Börnstein Editor of Springer-Verlag from 1993 through 1996. He has worked in several fields of solid state physics, namely electronic properties of defects, modulation spectroscopy, optical properties of semiconductors, surface physics, and scanning tunnelling microscopy (STM) in organic materials.

Claus Fischer

Formerly Institute of Solid State and Materials Research (IFW)
Dresden, Germany
A_C.FischerDD@t-online.de

Chapter 4.2

Claus Fischer recieved his PhD from the Technical University Dresden (Since his retirement in 2000 he continues to work as a foreign scientist of IFW in the field of high-T_c superconductors.) His last position at IFW was head of the Department of Superconducting Materials. The main areas of research were growth of metallic single crystals in particular of magnetic materials, developments of hard magnetic materials, of materials for thick film components of microelectronics and of low-T_c and high-T_c superconducting wires and tapes. Many activities were performed in cooperation with industrial manufacturers.

Günter Fuchs

Leibniz Institute for Solid State and
Materials Research (IFW) Dresden
Magnetism and Superconductivity in the
Institute of Metallic Materials
Dresden, Germany
fuchs@ifw-dresden.de
http://www.ifw-dresden.de/imw/21/

Chapter 4.2

Dr. Günter Fuchs studied physics at the Technical University of Dresden, Germany, and received his PhD in 1980 on the pinning mechanism in superconducting NbTi alloys. Since 1969 he has been at the Institute of Solid State and Materials Research (IFW) in Dresden. His activities are in superconductivity (HTSC, MgB_2, intermetallic borocarbides) and the applications of superconductors. He received the PASREG Award for outstanding scientific achievements in the field of bulk cuprate superconductors in high magnetic fields in 2003.

Frank Goodwin

International Lead Zinc Research
Organization, Inc.
Research Triangle Parc, NC, USA
fgoodwin@ilzro.org
http://www.ilzro.org/Contactus.htm

Chapter 3.1

Frank Goodwin received his Sc.D. from the Massachusetts Institute of Technology in 1979 and is responsible for all materials science research at International Lead Zinc Research Organization, Inc. where he has conceived and managed numerous projects on lead and zinc-containing products. These have included lead in acoustics, cable sheathing, nuclear waste management and specialty applications, together with zinc in coatings, castings and wrought forms.

Susana Gota-Goldmann

Commissariat à l'Energie Atomique (CEA)
Direction de la Recherche Technologique
(DRT)
Fontenay aux Roses, France
susana.gota-goldmann@cea.fr

Chapter 5.3

Dr. Susana Gota-Goldmann received her PhD in Materials Science form the Université Pierre et Marie Curie (Paris V) in 1993. After her PhD, she was engaged as a researcher in the Materials Science Division of the CEA (Commissariat à l'Energie Atomique, France). She has focused her scientific activity on the growth and characterisation of nanometric oxide layers with applications in spin electronics and photovoltaics. In parallel she has developed the use of synchrotron radiation techniques (X-ray absorption magnetic dicroism, photoemission, resonant reflectivity) for the study of oxide thin layers. Recently she has moved from fundamental to technological research. Dr. Gota-Goldmann is now working as a project manager at the scientific affairs direction of the Technology Research Division (CEA/DRT).

Sivaraman Guruswamy

University of Utah
Metallurgical Engineering
Salt Lake City, UT, USA
sguruswa@mines.utah.edu
http://www.mines.utah.edu/metallurgy/MML

Chapter 3.1

Dr. Guruswamy is a Professor of Metallurgical Engineering at the University of Utah. He obtained his Ph.D. degree in Metallurgical Engineering from the Ohio State University in 1984. He has made significant contributions in several areas including magnetic materials development, deformation of compound semiconductors, and lead alloys. His current work focuses on magnetostrictive materials and hybrid thermionic/thermoelectric thermal diodes.

Gagik G. Gurzadyan

Technical University of Munich
Institute for Physical and Theoretical
Chemistry
Garching, Germany
gurzadyan@ch.tum.de
http://zentrum.phys.chemie.tu-muenchen.de/gagik

Chapter 4.4

Gagik G. Gurzadyan, Ph.D., Dr. Sci., has extensive experience in nonlinear optics and crystals, laser photophysics and spectroscopy. He has authored several books including the Handbook of Nonlinear Optical Crystals published by Springer-Verlag. He worked in the Institute of Spectroscopy (USSR), CEA/Saclay (France), Max-Planck-Institute of Radiation Chemistry (Germany). At present he works at the Technical University of Munich with ultrafast lasers in the fields of nonlinear photochemistry of biomolecules and femtosecond spectroscopy.

Hideki Harada

High Tech Association Ltd.
Higashikaya, Fukaya,Saitama, Japan
khb16457@nifty.com
http://homepage1.nifty.com/JABM

Chapter 4.3

Dr. Hideki Harada is chief advisor of magnetic materials and their application and President of High Tech Association Ltd., Saitama, Japan. He is Chairman of the Japan Association of Bonded Magnet Industries (JABM) and received his Ph.D. in 1987 with a work on electrostatic ferrite materials. He worked in research and development of magnetic materials and cemented carbide tools at Hitachi Metals where he also was on the Board of Directors. He received the Japanese National Award for Industries Development Contribution.

Bernhard Holzapfel

Leibniz Institute for Solid State and Materials Research Dresden – Institute of Metallic Materials
Superconducting Materials
Dresden, Germany
B.Holzapfel@ifw-dresden.de
http://www.ifw-dresden.de/imw/26/

Chapter 4.2

Dr. Bernhard Holzapfel is head of the superconducting materials group at the Leibniz Institute for Solid State and Materials Research (IFW) Dresden, Germany. His main area of research is pulsed laser deposition of functional thin films and superconductivity. Currently he works on the development of HTSC high J_c coated conductors using ion beam assisted deposition or highly textured metal substrates. His work is supported by a number of national and European founded research projects.

Karl U. Kainer

GKSS Research Center Geesthacht
Institute for Materials Research
Geesthacht, Germany
karl.kainer@gkss.de
http://www.gkss.de

Chapter 3.1

Professor Kainer is director of Institute for Materials Research at GKSS-Research Center, Geesthacht and Professor of Materials Technology at the Technical University of Hamburg-Harburg. He obtained his Ph.D. in Materials Science at the Technical University of Clausthal in 1985 and his Habilitation in 1996. In 1988 he received the Japanese Government Research Award for Foreign Specialists. His current research activities are the development of new alloys and processes for magnesium materials.

Catrin Kammer

METALL – Intl. Journal for Metallurgy
Goslar, Germany
Kammer@metall-news.com
http://www.giesel-verlag.de

Chapter 3.1

Catrin Kammer received her Ph.D. in materials sciences from the Technical University Bergakademie Freiberg, Germany, in 1989. She has been working in the field of light metals and is author of several handbooks about aluminium and magnesium. She is working as author for the journal ALUMINIUM and is teaching in material sciences. Since 2001 she is editor-in-chief of the journal METALL, which deals with all non-ferrous metals.

Wolfram Knabl

Plansee AG
Technology Center
Reutte, Austria
wolfram.knabl@plansee.com
http://www.plansee.com

Chapter 3.1

Dr. Wolfram Knabl studied materials science at the Mining University of Leoben, Austria and received his Ph.D. at the Plansee AG focusing on the development of oxidation protective coatings for refractory metals. Between 1996 and 2002 he was responsible for the test laboratories at Plansee AG and since October 2002 he is working in the field of refractory metals, especially material and process development in the technology center of Plansee AG.

Alfred Koethe

Leibniz-Institut für Festkörper- und Werkstoffforschung
Institut für Metallische Werkstoffe (retired)
Dresden, Germany
alfred.koethe@web.de

Chapter 3.1

Dr. Alfred Koethe is physicist and professor of Materials Science. He retired in 2000 from his position as head of department in the Institute of Metallic Materials at the Leibniz Institute of Solid State and Materials Research in Dresden, Germany. His main research activities were in the fields of preparation and properties of ultrahigh-purity refractory metals and, especially, of steels (stainless steels, high strenght steels, thermomechanical treatment, microalloying, relations chemical composition/microstructure/properties).

Dieter Krause

Schott AG
Research and Technology-Development
Mainz, Germany
dieter.krause@schott.com

Chapter 3.4

Dieter Krause studied physics at the universities of Erlangen and Munich, Germany, where he received his Ph.D. for work on magnetism and metal physics. He was professor in Tehran, Iran, lecturer in Munich and Mainz, Germany. As scientist and director of Schott's corporate research and development centre he was involved in research on optical and mechanical properties of amorphous materials, thin films, and optical fibres. Now he is consultant, chief scientist, and the editor of the "Schott Series on Glass and Glass Ceramics – Science, Technology, and Applications" published by Springer.

Manfred D. Lechner

Universität Osnabrück
Institut für Chemie – Physikalische Chemie
Osnabrück, Germany
lechner@uni-osnabrueck.de
http://www.chemie.uni-osnabrueck.de/pc/index.html

Chapter 3.3

Professor Lechner has a PhD in chemistry from the University of Mainz, Germany. Since 1975 he is Professor of Physical Chemistry at the Institute of Chemistry of the University of Osnabrück, Germany. His scientific work concentrates on the physics and chemistry of polymers. In this area he is mainly working on the influence of high pressure on polymer systems, polymers for optical storage and waveguides as well as synthesis and properties of superabsorbers from renewable resources.

Gerhard Leichtfried

Plansee AG
Technology Center
Reutte, Austria
gerhard.leichtfried@plansee.com
http://www.plansee.com

Chapter 3.1

Dr. Gerhard Leichtfried received his Ph.D from the Montanuniversität Leoben and is qualified for lecturing in powder metallurgy. For 20 years he has been working in various senior positions for the Plansee Aktiengesellschaft, a company engaged in refractory metals, composite materials, cemented carbides and sintered iron and steels.

Werner Martienssen

Universität Frankfurt/Main
Physikalisches Institut
Frankfurt/Main, Germany
Martienssen@Physik.uni-frankfurt.de

Chapters 1.1, 1.2, 2.1, 4.1

Werner Martienssen studied physics and chemistry at the Universities of Würzburg and Göttingen, Germany. He obtained his Ph.D. in Physics with R.W. Pohl, Göttingen, and holds an honorary doctorate at the University of Dortmund. After a visiting-professorship at the Cornell University, Ithaca, USA in 1959 to 1960 he taught physics at the University of Stuttgart and since 1961 at the University of Frankfurt/Main. His main research fields are condensed matter physics, quantum optics and chaotic dynamics. Two of his former students and coworkers became Nobel-laureates in Physics, Gerd K. Binnig for the design of the scanning tunneling microscope in 1986 and Horst L. Störmer for the discovery of a new form of quantum-fluid with fractionally charged excitations in 1998. Werner Martienssen is a member of the Deutsche Akademie der Naturforscher Leopoldina, Halle and of the Akademie der Wissenschaften zu Göttingen. Since 1994 he is Editor-in-Chief of the data collection Landolt–Börnstein published by Springer, Heidelberg.

Toshio Mitsui

Osaka University
Takarazuka, Japan
t-mitsui@jttk.zaq.ne.jp

Chapter 4.5

Toshio Mitsui is an emeritus professor of Osaka University. He studied solid state physics and biophysics at Hokkaido University, Pennsylvania State University, Brookhaven National Laboratory, the Massachusetts Institute of Technology, Osaka University and Meiji University. He was the first to observe the ferroelectric domain structure in Rochelle salt with a polarization microscope. He proposed various theories on ferroelectric effects and biological molecular machines.

Manfred Müller

Dresden University of Technology
Institute of Materials Science
Dresden, Germany
m.mueller33@t-online.de

Chapter 4.3

Dr.-Ing. habil. Manfred Müller is a Professor emeritus of Special Materials at the Institute of Materials Science of the Dresden University of Technology. Before his retirement he was for many years head of department for special materials at the Central Institute for Solid State Physics and Materials Research of the Academy of Sciences in Dresden, Germany. His main field was the research and development of metallic materials with emphasis on special physical properties, such as soft and hard magnetic, electrical and thermoelastic properties. His last field of research was amorphous and nanocrystalline soft magnetic alloys. He is a member of the German Society of Materials Science (DGM) and was a member of the Advisary Board of DGM.

Sergei Pestov

Moscow State Academy of Fine Chemical Technology
Department of Inorganic Chemistry
Moscow, Russia
pestovsm@yandex.ru

Chapter 5.1

Dr. Pestov is a docent of the Inorganic Chemistry Department and a head of group on liquid crystals (LC) at the Moscow State Academy of Fine Chemical Technology. He earned his Ph.D. in physical chemistry in 1992. His research is focused on thermal analysis and thermodynamics of systems containing LC and physical properties of LC. He is an author of a Landolt–Börnstein volume and two books devoted to liquid crystals.

Günther Schlamp

Metallgesellschaft Ffm and Degussa
Demetron (retired)
Steinbach/Ts, Germany

Chapter 3.1

Günther Schlamp received his Ph.D. from the Johann-Wolfgang-Goethe University of Frankfurt/Main, Germany, in Physical Chemistry. His industrial activities in research include the development and production of refractory material coatings, high purity materials and parts for electronics, and sputter targets for the reflection-enhancing coating of glas. He has contributed to several Handbooks with repoprts on properties and applications of noble metals and their alloys.

Barbara Schüpp-Niewa

Leibniz-Institute for Solid State and Materials Research Dresden
Institute for Metallic Materials
Dresden, Germany
b.schuepp@ifw-dresden.de
http://www.ifw-dresden.de

Chapter 4.2

Barbara Schüpp-Niewa studied chemistry in Gießen and Dortmund where she received her Ph.D. in 1999. Since 2000 she has been a scientist at the Leibniz-Institute for Solid State and Materials Research Dresden with a focus on crystal structure investigations of oxometalates with superconducting or exciting magnetic ground states. Her current research activities include coated conductors.

Roland Stickler

University of Vienna
Department of Chemistry
Vienna, Austria
roland.stickler@univie.ac.at

Chapter 3.1

Professor Stickler received his master and Dr. degree from the Technical University in Vienna. From 1958 to 1972 he was manager of physical metallurgy with the Westinghouse Research Laboratory in Pittsburgh, Pa. In 1972 he accepted a full professorship at the University of Vienna heading a materials science group in the Institute of Physical Chemistry, and from 1988 he was head of this institute until his retirement as professor emeritus in 1998. He was involved in research and engineering work on superalloys, semiconductor materials and high melting point materials, investigating the relationship between microstructure and mechanical behavior, in particular fatigue and fracture mechanics properties. He was leader of a successful project on brazing under microgravity conditions in the Spacelab-Mission. Further activities included the participation in European COST projects, in particular as chairman of actions on powder metallurgy and light metals. He has authored and coauthored more than 250 publications in scientific journals and proceedings.

Pancho Tzankov

Max Born Institute for Nonlinear Optics and Short Pulse Spectroscopy
Berlin, Germany
tzankov@mbi-berlin.de
http://staff.mbi-berlin.de/tzankov/

Chapter 4.4

Pancho Tzankov studied laser physics at Sofia University, Bulgaria, and received his Ph.D. in physical chemistry from the Technical University of Munich, Germany. He is now a postdoctoral fellow at the Max Born Institute in Berlin, Germany. His research activities involve development of new nonlinear optical parametric sources of ultrashort pulses and their application for time-resolved spectroscopy.

Volkmar Vill

University of Hamburg
Department of Chemistry, Institute of Organic Chemistry
Hamburg, Germany
vill@chemie.uni-hamburg.de
http://liqcryst.chemie.uni-hamburg.de/

Chapter 5.1

Professor Volkmar Vill received his Diploma in Chemistry in 1986, his Diploma in Physics in 1988 and his Ph.D. in Chemistry in 1990 from the University of Münster, Germany. In 1997 he earned his Habilitation in Organic Chemistry from the University of Hamburg where he is Professor of Organic Chemistry since 2002. He is the author of the LiqCryst – Database of Liquid Crystals and the Editor of the Handbook of Liquid Crystals, of Landolt–Börnstein, Organic Index, and Vol. VIII/5a, Physical Properties of Liquid Crystals.

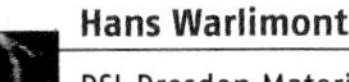

Hans Warlimont

DSL Dresden Material-Innovation GmbH
Dresden, Germany
warlimont@ifw-dresden.de

Chapters 3.1, 3.2, 4.2, 4.3

Hans Warlimont is a physical metallurgist and has worked on numerous topics in several research institutions and industrial companies. Among them were the Max-Planck-Institute of Metals Research, Stuttgart, and Vacuumschmelze, Hanau. He was Scientific Director of the Leibniz-Institute of Solid State and Materials Research Dresden and Professor of Materials Science at Dresden University of Technology. Recently he has established DSL Dresden Material-Innovation GmbH to industrialise his invention of electroformed battery grids.

Acknowledgements

2.1 The Elements

by Werner Martienssen

We thank Dr. G. Leichtfried, Plansee AG, A-6600 Reutte/Tirol for recently determined new data on the refractory metals Nb, Ta, and Mo, W.

4.1 Semiconductors

by Werner Martienssen

In selecting the "most important information" from the huge data collection in Landolt–Börnstein, the author found great help in the new *Semiconductors: Data Handbook* [1]. Again, the data in this Springer Handbook of Condensed Matter and Materials Data represent only a small fraction of the information given in *Semiconductors: Data Handbook*, which is about 700 pages long. I am much indebted to my colleague O. Madelung for kindly presenting me the manuscript of that Handbook prior to publication.

[1] O. Madelung (Ed.): *Semiconductors: Data Handbook*, 3rd Edn. (Springer, Berlin, Heidelberg 2004)

4.5 Ferroelectrics and Antiferroelectrics

by Toshio Mitsui

The author of this subchapter thanks the coauthors of LB III/36 for their helpful discussions and suggestions. Especially, he is much indebted to Prof. K. Deguchi for his kind support throughout the preparation of the manuscript.

Contents

List of Abbreviations 19

第1册 通用表和元素（本册）

Part 1 General Tables

1 **The Fundamental Constants**
Werner Martienssen 3
1.1 What are the Fundamental Constants
and Who Takes Care of Them? 3
1.2 The CODATA Recommended Values of the Fundamental Constants 4
References 9

2 **The International System of Units (SI), Physical Quantities, and Their Dimensions**
Werner Martienssen 11
2.1 The International System of Units (SI) 11
2.2 Physical Quantities 12
2.3 The SI Base Units 13
2.4 The SI Derived Units 16
2.5 Decimal Multiples and Submultiples of SI Units 19
2.6 Units Outside the SI 20
2.7 Some Energy Equivalents 24
References 25

3 **Rudiments of Crystallography**
Wolf Assmus, Stefan Brühne 27
3.1 Crystalline Materials 28
3.2 Disorder 38
3.3 Amorphous Materials 39
3.4 Methods for Investigating Crystallographic Structure 39
References 41

Part 2 The Elements

1 **The Elements**
Werner Martienssen 45
1.1 Introduction 45
1.2 Description of Properties Tabulated 46
1.3 Sources 49
1.4 Tables of the Elements in Different Orders 49
1.5 Data 54
References 158

第2册 材料类：金属材料

Part 3 Classes of Materials

1 **Metals**
Frank Goodwin, Sivaraman Guruswamy, Karl U. Kainer, Catrin Kammer, Wolfram Knabl, Alfred Koethe, Gerhard Leichtfried, Günther Schlamp, Roland Stickler, Hans Warlimont 161
1.1 Magnesium and Magnesium Alloys 162
1.2 Aluminium and Aluminium Alloys 171
1.3 Titanium and Titanium Alloys 206
1.4 Zirconium and Zirconium Alloys 217
1.5 Iron and Steels 221
1.6 Cobalt and Cobalt Alloys 272
1.7 Nickel and Nickel Alloys 279
1.8 Copper and Copper Alloys 296
1.9 Refractory Metals and Alloys 303
1.10 Noble Metals and Noble Metal Alloys 329
1.11 Lead and Lead Alloys 407
References 422

第3册 材料类：非金属材料

2 **Ceramics**
Hans Warlimont 431
2.1 Traditional Ceramics and Cements 432
2.2 Silicate Ceramics 433
2.3 Refractory Ceramics 437
2.4 Oxide Ceramics 437
2.5 Non-Oxide Ceramics 451
References 476

3 **Polymers**
Manfred D. Lechner 477
3.1 Structural Units of Polymers 480
3.2 Abbreviations 482
3.3 Tables and Figures 483
References 522

4 **Glasses**
Dieter Krause 523
4.1 Properties of Glasses – General Comments 526
4.2 Composition and Properties of Glasses 527
4.3 Flat Glass and Hollowware 528
4.4 Technical Specialty Glasses 530
4.5 Optical Glasses 543
4.6 Vitreous Silica 556
4.7 Glass-Ceramics 558

4.8 Glasses for Miscellaneous Applications 559
References 572

第4册 功能材料：半导体和超导体

Part 4 Functional Materials

1 **Semiconductors**
Werner Martienssen 575
1.1 Group IV Semiconductors and IV–IV Compounds 578
1.2 III–V Compounds 604
1.3 II–VI Compounds 652
References 691

2 **Superconductors**
Claus Fischer, Günter Fuchs, Bernhard Holzapfel, Barbara Schüpp-Niewa, Hans Warlimont 695
2.1 Metallic Superconductors 696
2.2 Non-Metallic Superconductors 711
References 749

第5册 功能材料：磁性材料、电介质、铁电体和反铁电体

3 **Magnetic Materials**
Hideki Harada, Manfred Müller, Hans Warlimont 755
3.1 Basic Magnetic Properties 755
3.2 Soft Magnetic Alloys 758
3.3 Hard Magnetic Alloys 794
3.4 Magnetic Oxides 811
References 814

4 **Dielectrics and Electrooptics**
Gagik G. Gurzadyan, Pancho Tzankov 817
4.1 Dielectric Materials: Low-Frequency Properties 822
4.2 Optical Materials: High-Frequency Properties 824
4.3 Guidelines for Use of Tables 826
4.4 Tables of Numerical Data for Dielectrics and Electrooptics 828
References 890

5 **Ferroelectrics and Antiferroelectrics**
Toshio Mitsui 903
5.1 Definition of Ferroelectrics and Antiferroelectrics 903
5.2 Survey of Research on Ferroelectrics 904
5.3 Classification of Ferroelectrics 906
5.4 Physical Properties of 43 Representative Ferroelectrics 912
References 936

第6册 特种结构

Part 5 Special Structures

1 Liquid Crystals
Sergei Pestov, Volkmar Vill 941
1.1 Liquid Crystalline State 941
1.2 Physical Properties of the Most Common Liquid Crystalline Substances 946
1.3 Physical Properties of Some Liquid Crystalline Mixtures 975
References 977

2 The Physics of Solid Surfaces
Gianfranco Chiarotti 979
2.1 The Structure of Ideal Surfaces 979
2.2 Surface Reconstruction and Relaxation 986
2.3 Electronic Structure of Surfaces 996
2.4 Surface Phonons 1012
2.5 The Space Charge Layer at the Surface of a Semiconductor 1020
2.6 Most Frequently Used Acronyms 1026
References 1029

3 Mesoscopic and Nanostructured Materials
Fabrice Charra, Susana Gota-Goldmann 1031
3.1 Introduction and Survey 1031
3.2 Electronic Structure and Spectroscopy 1035
3.3 Electromagnetic Confinement 1044
3.4 Magnetic Nanostructures 1048
3.5 Preparation Techniques 1063
References 1066

Subject Index
Periodic Table of the Elements
Most Frequently Used Fundamental Constants

List of Abbreviations

2D-BZ 2-dimensional Brillouin zone
2P-PES 2-photon photoemission spectroscopy

A

AES Auger electron spectroscopy
AFM atomic force microscope
AISI American Iron and Steel Institute
APS appearance potential spectroscopy
ARUPS angle-resolved ultraviolet photoemission spectroscopy
ARXPS angle-resolved X-ray photoemission spectroscopy
ASTM American Society for Testing and Materials
ATR attenuated total reflection

B

BBZ bulk Brillouin zone
BIPM Bureau International des Poids et Mesures
BZ Brillouin zone

C

CB conduction band
CBM conduction band minimum
CISS collision ion scattering spectroscopy
CITS current imaging tunneling spectroscopy
CMOS complementary metal–oxide–semiconductor
CODATA Committee on Data for Science and Technology
CVD chemical vapour deposition

D

DFB distributed-feedback
DFG difference frequency generation
DOS density of states
DSC differential scanning calorimetry
DTA differential thermal analysis

E

EB electron-beam melting
ECS electron capture spectroscopy
EELS electron-energy loss spectroscopy
ELEED elastic low-energy electron diffraction
ESD electron-stimulated desorption
EXAFS extended X-ray absorption fine structure

F

FEM field emission microscope/microscopy
FIM field ion microscope/microscopy

G

GMR giant magnetoresistance

H

HAS helium atom scattering
HATOF helium atom time-of-flight spectroscopy
HB Brinell hardness number
HEED high-energy electron diffraction
HEIS high-energy ion scattering/high-energy ion scattering spectroscopy
HK Knoop hardness
HOPG highly oriented pyrolytic graphite
HPDC high-pressure die casting
HR-EELS high-resolution electron energy loss spectroscopy
HR-LEED high-resolution LEED
HR-RHEED high-resolution RHEED
HREELS high-resolution electron energy loss spectroscopy
HRTEM high-resolution transition electron microscopy
HT high temperature
HTSC high-temperature superconductor
HV Vicker's Hardness

I

IACS International Annealed Copper Standard
IB ion bombardment
IBAD ion-beam-assisted deposition
ICISS impact ion scattering spectroscopy
ICSU International Council of the Scientific Unions
IPE inverse photoemission
IPES inverse photoemission spectroscopy
ISO International Organization for Standardization
ISS ion scattering spectroscopy
IUPAC International Union of Pure and Applied Chemistry

J

JDOS joint density of states

K

KRIPES K-resolved inverse photoelectron spectroscopy

L

LAPW linearized augmented-plane-wave method
LB Langmuir–Blodgett
LCM liquid crystal material
LCP liquid crystal polymer
LCs liquid crystals
LDA local-density approximation
LDOS local density of states
LEED low-energy electron diffraction
LEIS low-energy ion scattering/low-energy ion scattering spectroscopy
LPE liquid phase epitaxy

M

MBE molecular-beam epitaxy
MD molecular dynamics
MEED medium-energy electron diffraction
MEIS medium-energy ion scattering/medium-energy ion scattering spectroscopy
MFM magnetic force microscopy
ML monolayer
MOCVD metal-organic chemical vapor deposition
MOKE magneto-optical Kerr effect
MOSFET MOS field-effect transistor
MQW multiple quantum well

N

NICISS neutral impact collision ion scattering spectroscopy
NIMs National Institutes for Metrology

O

OPO optical parametric oscillation

P

PDS photothermal displacement spectroscopy
PED photoelectron diffraction
PES photoemission spectroscopy
PLAP pulsed laser atom probe
PLD pulsed laser deposition
PSZ stabilized zirconia
PZT piezoelectric material

R

RAS reflectance anisotropy spectroscopy
RE rare earth
REM reflection electron microscope/microscopy
RHEED reflection high-energy electron diffraction
RIE reactive ion etching
RPA random-phase approximation
RT room temperature
RTP room temperaure and standard pressure

S

SAM self-assembled monolayer
SAM scanning Auger microscope/microscopy
SARS scattering and recoiling ion spectroscopy
SAW surface acoustic wave
SBZ surface Brillouin zone
SCLS surface core level shift
SDR surface differential reflectivity
SEM scanning electron microscope
SEXAFS surface-sensitive EXAFS
SFG sum frequency generation
SH second harmonic
SHG second-harmonic generation
SI Système International d'Unités
SIMS secondary-ion mass spectroscopy
SNR signal-to-noise ratio
SPARPES spin polarized angle-resolved photoemission spectroscopy
SPIPES spin-polarized inverse photoemission spectroscopy
SPLEED spin-polarized
SPV surface photovoltage spectroscopy
SQUIDS superconducting quantum interference devices
SS surface state
STM scanning tunneling microscope/microscopy
STS scanning tunneling spectroscopy
SXRD surface X-ray diffraction

T

TAFF thermally activated flux flow
TEM transmission electron microscope/microscopy
TFT thin-film transistor
TMR tunnel magnetoresistance
TMT thermomechanical treatment
TOF time of flight
TOM torsion oscillation magnetometry

TRS truncation rod scattering
TTT time-temperature-transformation

U

UHV ultra-high vacuum
UPS ultraviolet photoemission spectroscopy
UV ultraviolet

V

VBM valence band maximum
VLEED very low-energy electron diffraction

X

XPS X-ray photoemission spectroscopy

Part 1

General Ta

Part 1 General Tables

1 **The Fundamental Constants**
Werner Martienssen, Frankfurt/Main, Germany

2 **The International System of Units (SI), Physical Quantities, and Their Dimensions**
Werner Martienssen, Frankfurt/Main, Germany

3 **Rudiments of Crystallography**
Wolf Assmus, Frankfurt am Main, Germany
Stefan Brühne, Frankfurt am Main, Germany

1.1. The Fundamental Constants

In the quantitative description of physical phenomena and physical relationships, we find constant parameters which appear to be independent of the scale of the phenomena, independent of the place where the phenomena happen, and independent of the time when the phenomena is observed. These parameters are called fundamental constants. In Sect. 1.1.1, we give a qualitative description of these basic parameters and explain how "recommended values" for the numerical values of the fundamental constants are found. In Sect. 1.1.2, we present tables of the most recently determined recommended numerical values for a large number of those fundamental constants which play a role in solid-state physics and chemistry and in materials science.

1.1.1 What are the Fundamental Constants and Who Takes Care of Them?

The fundamental constants are constant parameters in the laws of nature. They determine the size and strength of the phenomena in the natural and technological worlds. We conclude from observation that the numerical values of the fundamental constants are independent of space and time; at least, we can say that if there is any dependence of the fundamental constants on space and time, then this dependence must be an extremely weak one. Also, we observe that the numerical values are independent of the scale of the phenomena observed; for example, they seem to be the same in astrophysics and in atomic physics. In addition, the numerical values are quite independent of the environmental conditions. So we have confidence in the idea that the numerical values of the fundamental constants form a set of numbers which are the same everywhere in the world, and which have been the same in the past and will be the same in the future. Whereas the properties of all material objects in nature are more or less subject to continuous change, the fundamental constants seem to represent a constituent of the world which is absolutely permanent.

On the basis of this expected invariance of the fundamental constants in space and time, it appears reasonable to relate the units of measurement for physical quantities to fundamental constants as far as possible. This would guarantee that also the units of measurement become independent of space and time and of environmental conditions. Within the frame work of the International System of Units (Système International d'Unités, abbreviated to SI), the International Committee for Weights and Measures (Comité International des Poids et Mesures, CIPM) has succeeded in relating a large number of units of measurement for physical quantities to the numerical values of selected fundamental constants; however, several units for physical quantities are still represented by prototypes. For example, the unit of length 1 meter, is defined as the distance light travels in vacuum during a fixed time; so the unit of length is related to the fundamental constant c, i. e. the speed of light, and the unit of time, 1 second. The unit of mass, 1 kilogram, however, is still represented by a prototype, the mass of a metal cylinder made of a platinum–iridium alloy, which is carefully stored at the International Office for Weights and Measures (Bureau International des Poids et Mesures, BIPM), at Sèvres near Paris. In a few years, however, it might become possible also to relate the unit of mass to one or more fundamental constants.

The fundamental constants play an important role in basic physics as well as in applied physics and technology; in fact, they have a key function in the development of a system of reproducible and unchanging units for physical quantities. Nevertheless, there is, at present, no theory which would allow us to calculate the numerical values of the fundamental constants. Therefore, national institutes for metrology (NIMs), together with research institutes and university laboratories, are making efforts worldwide to determine the fundamental constants experimentally with the greatest possible accuracy and reliability. This, of course, is a continuous process, with hundreds of new publications every year.

The Committee on Data for Science and Technology (CODATA), established in 1966 as an interdisciplinary, international committee of the International Council of the Scientific Unions (ICSU), has taken the responsibility for improving the quality, reliability, processing, management, and accessibility of data of importance to science and technology. The CODATA task group on fundamental constants, established in 1969, has taken on the job of periodically providing the scientific and technological community with a self-consistent set of internationally recommended values of the fundamental constants based on all relevant data available at given points in time.

What is the meaning of "recommended values" of the fundamental constants?

Many fundamental constants are not independent of one another; they are related to one another by equations which allow one to calculate a numerical value for one particular constant from the numerical values of other constants. In consequence, the numerical value of a constant can be determined either by measuring it directly or by calculating it from the measured values of other constants related to it. In addition, there are usually several different experimental methods for measuring the value of any particular fundamental constant. This allows one to compute an adjustment on the basis of a least-squares fit to the whole set of experimental data in order to determine a set of best-fitting fundamental constants from the large set of all experimental data. Such an adjustment is done today about every four years by the CODATA task group mentioned above. The resulting set of best-fit values is then called the "CODATA recommended values of the fundamental constants" based on the adjustment of the appropriate year.

The tables in Sect. 1.1.2 show the CODATA recommended values of the fundamental constants of science and technology based on the 2002 adjustment. This adjustment takes into account all data that became available before 31 December 2002. A detailed description of the adjustment has been published by *Mohr* and *Taylor* of the National Institute of Standards and Technology, Gaithersburg, in [1.1].

The editors of this Handbook, H. Warlimont and W. Martienssen, would like to express their sincere thanks to Mohr and Taylor for kindly putting their data at our disposal prior to publication in Reviews of Modern Physics. These data first became available in December 2003 on the web site of the NIST fundamental constants data center [1.2].

1.1.2 The CODATA Recommended Values of the Fundamental Constants

1.1.2.1 The Most Frequently Used Fundamental Constants

Tables 1.1-1 – 1.1-9 list the CODATA recommended values of the fundamental constants based on the 2002 adjustment.

Table 1.1-1 Brief list of the most frequently used fundamental constants

Quantity	Symbol and relation	Numerical value	Units	Relative standard uncertainty
Speed of light in vacuum	c	299 792 458	m/s	Fixed by definition
Magnetic constant	$\mu_0 = 4\pi \times 10^{-7}$	$12.566370614\ldots \times 10^{-7}$	N/A^2	Fixed by definition
Electric constant	$\varepsilon_0 = 1/(\mu_0 c^2)$	$8.854187817\ldots \times 10^{-12}$	F/m	Fixed by definition
Newtonian constant of gravitation	G	$6.6742(10) \times 10^{-11}$	$m^3/(kg\ s^2)$	1.5×10^{-4}
Planck constant	h	$4.13566743(35) \times 10^{-15}$	eV s	8.5×10^{-8}
Reduced Planck constant	$\hbar = h/2\pi$	$6.58211915(56) \times 10^{-16}$	eV s	8.5×10^{-8}
Elementary charge	e	$1.60217653(14) \times 10^{-19}$	C	8.5×10^{-8}

Table 1.1-1 Brief list of the most frequently used fundamental constants, cont.

Quantity	Symbol and relation	Numerical value	Units	Relative standard uncertainty
Fine-structure constant	$\alpha = (1/4\pi\varepsilon_0)(e^2/\hbar c)$	$7.297352568(24) \times 10^{-3}$		3.3×10^{-9}
Magnetic flux quantum	$\Phi_0 = h/2e$	$2.06783372(18) \times 10^{-15}$	Wb	8.5×10^{-8}
Conductance quantum	$G_0 = 2e^2/h$	$7.748091733(26) \times 10^{-5}$	S	3.3×10^{-9}
Rydberg constant	$R_\infty = \alpha^2 m_e c/2h$	10 973 731.568525(73)	1/m	6.6×10^{-12}
Electron mass	m_e	$9.1093826(16) \times 10^{-31}$	kg	1.7×10^{-7}
Proton mass	m_p	$1.67262171(29) \times 10^{-27}$	kg	1.7×10^{-7}
Proton–electron mass ratio	m_p/m_e	1836.15267261(85)		4.6×10^{-10}
Avogadro number	N_A, L	$6.0221415(10) \times 10^{23}$		1.7×10^{-7}
Faraday constant	$F = N_A e$	96 485.3383(83)	C	8.6×10^{-8}
Molar gas constant	R	8.314472(15)	J/K	1.7×10^{-6}
Boltzmann constant	$k = R/N_A$	$1.3806505(24) \times 10^{-23}$ $8.617343(15) \times 10^{-5}$	J/K eV/K	1.8×10^{-6} 1.8×10^{-6}
Stefan–Boltzmann constant	$\sigma = (\pi^2/60)(k^4/(\hbar^3 c^2))$	$5.670400(40) \times 10^{-8}$	$W/(m^2\,K^4)$	7.0×10^{-6}

1.1.2.2 Detailed Lists of the Fundamental Constants in Different Fields of Application

Table 1.1-2 Universal constants

Quantity	Symbol and relation	Numerical value	Units	Relative standard uncertainty
Speed of light in vacuum	c	299 792 458	m/s	Fixed by definition
Magnetic constant	$\mu_0 = 4\pi \times 10^{-7}$	$12.566370614\ldots \times 10^{-7}$	N/A^2	Fixed by definition
Electric constant	$\varepsilon_0 = 1/(\mu_0 c^2)$	$8.854187817\ldots \times 10^{-12}$	F/m	Fixed by definition
Characteristic impedance of vacuum	$Z_0 = (\mu_0/\varepsilon_0)^{1/2} = \mu_0 c$	376.730313461 ...	Ω	Fixed by definition
Newtonian constant of gravitation	G	$6.6742(10) \times 10^{-11}$	$m^3/(kg\,s^2)$	1.5×10^{-4}
Reduced Planck constant	$G/\hbar c$	$6.7087(10) \times 10^{-39}$	$(GeV/c^2)^{-2}$	1.5×10^{-4}
Planck constant	h	$6.6260693(11) \times 10^{-34}$ $4.13566743(35) \times 10^{-15}$	J s eV s	1.7×10^{-7} 8.5×10^{-8}
(Ratio)	$\hbar = h/2\pi$	$1.05457168(18) \times 10^{-34}$ $6.58211915(56) \times 10^{-16}$	J s eV s	1.7×10^{-7} 8.5×10^{-8}
(Product)	$\hbar c$	197.326968(17)	MeV fm	8.5×10^{-8}
(Product)	$c_1 = 2\pi h c^2$	$3.74177138(64) \times 10^{-16}$	$W\,m^2$	1.7×10^{-7}
(Product)	$(1/\pi)c_1 = 2hc^2$	$1.19104282(20) \times 10^{-16}$	$W\,m^2/sr$	1.7×10^{-7}
(Product)	$c_2 = h(c/k)$	$1.4387752(25) \times 10^{-2}$	m K	1.7×10^{-6}
Stefan–Boltzmann constant	$\sigma = (\pi^2/60)(k^4/(\hbar^3 c^2))$	$5.670400(40) \times 10^{-8}$	$W/(m^2\,K^4)$	7.0×10^{-6}
Wien displacement law constant	$b = \lambda_{max} T = c_2/4.965114231$	$2.8977685(51) \times 10^{-3}$	m K	1.7×10^{-6}
Planck mass	$m_P = (\hbar c/G)^{1/2}$	$2.17645(16) \times 10^{-8}$	kg	7.5×10^{-5}
Planck temperature	$T_P = (1/k)(\hbar c^5/G)^{1/2}$	$1.41679(11) \times 10^{32}$	K	7.5×10^{-5}
Planck length	$l_P = \hbar/m_P c$ $= (\hbar G/c^3)^{1/2}$	$1.61624(12) \times 10^{-35}$	m	7.5×10^{-5}
Planck time	$t_P = l_P/c = (\hbar G/c^5)^{1/2}$	$5.39121(40) \times 10^{-44}$	s	7.5×10^{-5}

Table 1.1-3 Electromagnetic constants

Quantity	Symbol and relation	Numerical value	Units	Relative standard uncertainty
Elementary charge	e	$1.60217653(14)\times10^{-19}$	C	8.5×10^{-8}
(Ratio)	e/h	$2.41798940(21)\times10^{14}$	A/J	8.5×10^{-8}
Fine-structure constant	$\alpha=(1/4\pi\varepsilon_0)(e^2/\hbar c)$	$7.297352568(24)\times10^{-3}$		3.3×10^{-9}
Inverse fine-structure constant	$1/\alpha$	137.03599911(46)		3.3×10^{-9}
Magnetic flux quantum	$\Phi_0=h/2e$	$2.06783372(18)\times10^{-15}$	Wb	8.5×10^{-8}
Conductance quantum	$G_0=2e^2/h$	$7.748091733(26)\times10^{-5}$	S	3.3×10^{-9}
Inverse of conductance quantum	$1/G_0$	12 906.403725(43)	Ω	3.3×10^{-9}
Josephson constant [a]	$K_J=2e/h$	$483\,597.879(41)\times10^{9}$	Hz/V	8.5×10^{-8}
von Klitzing constant [b]	$R_K=h/e^2=\mu_0 c/2\alpha$	25 812.807449(86)	Ω	3.3×10^{-9}
Bohr magneton	$\mu_B=e\hbar/2m_e$	$927.400949(80)\times10^{-26}$ $5.788381804(39)\times10^{-5}$	J/T eV/T	8.6×10^{-8} 6.7×10^{-9}
(Ratio)	μ_B/h	$13.9962458(12)\times10^{9}$	Hz/T	8.6×10^{-8}
(Ratio)	μ_B/hc	46.6864507(40)	1/(m T)	8.6×10^{-8}
(Ratio)	μ_B/k	0.6717131(12)	K/T	1.8×10^{-6}
Nuclear magneton	$\mu_N=e\hbar/2m_p$	$5.05078343(43)\times10^{-27}$ $3.152451259(21)\times10^{-8}$	J/T eV/T	8.6×10^{-8} 6.7×10^{-9}
(Ratio)	μ_N/h	7.62259371(65)	MHz/T	8.6×10^{-8}
(Ratio)	μ_N/hc	$2.54262358(22)\times10^{-2}$	1/(m T)	8.6×10^{-8}
(Ratio)	μ_N/k	$3.6582637(64)\times10^{-4}$	K/T	1.8×10^{-6}

[a] See Table 1.2-16 for the conventional value adopted internationally for realizing representations of the volt using the Josephson effect.
[b] See Table 1.2-16 for the conventional value adopted internationally for realizing representations of the ohm using the quantum Hall effect.

Table 1.1-4 Thermodynamic constants

Quantity	Symbol and relation	Numerical value	Units	Relative standard uncertainty
Avogadro number	N_A, L	$6.0221415(10)\times10^{23}$		1.7×10^{-7}
Atomic mass constant	$u=(1/12)m(^{12}C)$ $=(1/N_A)\times10^{-3}$ kg	$1.66053886(28)\times10^{-27}$	kg	1.7×10^{-7}
Energy equivalent of atomic mass constant	$m_u c^2$	$1.49241790(26)\times10^{-10}$ 931.494043(80)	J MeV	1.7×10^{-7} 8.6×10^{-8}
Faraday constant	$F=N_A e$	96 485.3383(83)	C	8.6×10^{-8}
Molar Planck constant	$N_A h$	$3.990312716(27)\times10^{-10}$	J s	6.7×10^{-9}
(Product)	$N_A hc$	0.11962656572(80)	J m	6.7×10^{-9}
Molar gas constant	R	8.314472(15)	J/K	1.7×10^{-6}
Boltzmann constant	$k=R/N_A$	$1.3806505(24)\times10^{-23}$ $8.617343(15)\times10^{-5}$	J/K eV/K	1.8×10^{-6} 1.8×10^{-6}
(Ratio)	k/h	$2.0836644(36)\times10^{10}$	Hz/K	1.7×10^{-6}
(Ratio)	k/hc	69.50356(12)	1/(m K)	1.7×10^{-6}
Molar volume of ideal gas at STP	$V_m=RT/p$ at $T=273.15$ K and $p=101.325$ kPa	$22.413996(39)\times10^{-3}$	m^3	1.7×10^{-6}
Loschmidt constant	$n_0=N_A/V_m$	$2.6867773(47)\times10^{25}$	$1/m^3$	1.8×10^{-6}
Stefan–Boltzmann constant	$\sigma=(\pi^2/60)(k^4/(\hbar^3c^2))$	$5.670400(40)\times10^{-8}$	$W/(m^2\,K^4)$	7.0×10^{-6}
Wien displacement law constant	$b=\lambda_{max}T=c_2/4.965114231$	$2.8977685(51)\times10^{-3}$	m K	1.7×10^{-6}

1.1.2.3 Constants from Atomic Physics and Particle Physics

Table 1.1-5 Constants from atomic physics

Quantity	Symbol and relation	Numerical value	Units	Relative standard uncertainty
Rydberg constant	$R_\infty = \alpha^2 m_e c/2h$	10 973 731.568525(73)	1/m	6.6×10^{-12}
(Product)	$R_\infty c$	$3.289841960360(22) \times 10^{15}$	Hz	6.6×10^{-12}
(Product)	$R_\infty hc$	$2.17987209(37) \times 10^{-18}$ 13.6056923(12)	J eV	1.7×10^{-7} 8.5×10^{-8}
Bohr radius	$a_0 = \alpha/4\pi R_\infty$ $= 4\pi\varepsilon_0\hbar^2/m_e e^2$	$0.5291772108(18) \times 10^{-10}$	m	3.3×10^{-9}
Hartree energy	$E_H = e^2/4\pi\varepsilon_0 a_0$ $= 2R_\infty hc = \alpha^2 m_e c^2$	$4.35974417(75) \times 10^{-18}$ 27.2113845(23)	J eV	1.7×10^{-7} 8.5×10^{-8}
Quantum of circulation	$h/2m_e$	$3.636947550(24) \times 10^{-4}$	m^2/s	6.7×10^{-9}
(Product)	h/m_e	$7.273895101(48) \times 10^{-4}$	m^2/s	6.7×10^{-9}

Table 1.1-6 Properties of the electron

Quantity	Symbol and relation	Numerical value	Units	Relative standard uncertainty
Electron mass	m_e	$9.1093826(16) \times 10^{-31}$ $5.4857990945(24) \times 10^{-4}$	kg u	1.7×10^{-7} 4.4×10^{-10}
Energy equivalent of electron mass	$m_e c^2$	$8.1871047(14) \times 10^{-14}$ 0.510998918(44)	J MeV	1.7×10^{-7} 8.6×10^{-8}
Electron–proton mass ratio	m_e/m_p	$5.4461702173(25) \times 10^{-4}$		4.6×10^{-10}
Electron–neutron mass ratio	m_e/m_n	$5.4386734481(38) \times 10^{-4}$		7.0×10^{-10}
Electron–muon mass ratio	m_e/m_μ	$4.83633167(13) \times 10^{-3}$		2.6×10^{-8}
Electron molar mass	$M(e) = N_A m_e$	$5.4857990945(24) \times 10^{-7}$	kg	4.4×10^{-10}
Charge-to-mass ratio	$-e/m_e$	$-1.75882012(15) \times 10^{11}$	C/kg	8.6×10^{-8}
Compton wavelength	$\lambda_C = h/m_e c$	$2.426310238(16) \times 10^{-12}$	m	6.7×10^{-9}
(Ratio)	$\lambda_C/2\pi = \alpha a_0 = \alpha^2/(4\pi R_\infty)$	$386.1592678(26) \times 10^{-15}$	m	6.7×10^{-9}
Classical electron radius	$r_e = \alpha^2 a_0$	$2.817940325(28) \times 10^{-15}$	m	1.0×10^{-8}
Thomson cross section	$\sigma_e = (8\pi/3) r_e^2$	$0.665245873(13) \times 10^{-28}$	m^2	2.0×10^{-8}
Magnetic moment	μ_e	$-928.476412(80) \times 10^{-26}$	J/T	8.6×10^{-8}
Ratio of magnetic moment to Bohr magneton	μ_e/μ_B	$-1.0011596521859(38)$		3.8×10^{-12}
Ratio of magnetic moment to nuclear magneton	μ_e/μ_N	$-1838.28197107(85)$		4.6×10^{-10}
Ratio of magnetic moment to proton magnetic moment	μ_e/μ_p	$-658.2106862(66)$		1.0×10^{-8}
Ratio of magnetic moment to neutron magnetic moment	μ_e/μ_n	960.92050(23)		2.4×10^{-7}
Electron magnetic-moment anomaly	$a_e = \|\mu_e\|/(\mu_B - 1)$	$1.1596521859(38) \times 10^{-3}$		3.2×10^{-9}
g-factor	$g_e = -2(1 + a_e)$	$-2.0023193043718(75)$		3.8×10^{-12}
Gyromagnetic ratio	$\gamma_e = 2\|\mu_e\|/\hbar$	$1.76085974(15) \times 10^{11}$	1/(s T)	8.6×10^{-8}
(Ratio)	$\gamma_e/2\pi$	28 024.9532(24)	MHz/T	8.6×10^{-8}

Table 1.1-7 Properties of the proton

Quantity	Symbol and relation	Numerical value	Units	Relative standard uncertainty
Proton mass	m_p	$1.67262171(29) \times 10^{-27}$ $1.00727646688(13)$	kg u	1.7×10^{-7} 1.3×10^{-10}
Energy equivalent of proton mass	$m_p c^2$	$1.50327743(26) \times 10^{-10}$ $938.272029(80)$	J MeV	1.7×10^{-7} 8.6×10^{-8}
Proton–electron mass ratio	m_p/m_e	$1836.15267261(85)$		4.6×10^{-10}
Proton–neutron mass ratio	m_p/m_n	$0.99862347872(58)$		5.8×10^{-10}
Proton molar mass	$M(p) = N_A m_p$	$1.00727646688(13) \times 10^{-3}$	kg	1.3×10^{-10}
Charge-to-mass ratio	e/m_p	$9.57883376(82) \times 10^{7}$	C/kg	8.6×10^{-8}
Compton wavelength	$\lambda_{C,p} = h/m_p c$	$1.3214098555(88) \times 10^{-15}$	m	6.7×10^{-9}
(Ratio)	$(1/2\pi)\lambda_{C,p}$	$0.2103089104(14) \times 10^{-15}$	m	6.7×10^{-9}
rms charge radius	R_p	$0.8750(68) \times 10^{-15}$	m	7.8×10^{-3}
Magnetic moment	μ_p	$1.41060671(12) \times 10^{-26}$	J/T	8.7×10^{-8}
Ratio of magnetic moment to Bohr magneton	μ_p/μ_B	$1.521032206(15) \times 10^{-3}$		1.0×10^{-8}
Ratio of magnetic moment to nuclear magneton	μ_p/μ_N	$2.792847351(28)$		1.0×10^{-8}
Ratio of magnetic moment to neutron magnetic moment	μ_p/μ_n	$-1.45989805(34)$		2.4×10^{-7}
g-factor	$g_p = 2\mu_p/\mu_N$	$5.585694701(56)$		1.0×10^{-8}
Gyromagnetic ratio	$\gamma_p = 2\mu_p/\hbar$	$2.67522205(23) \times 10^{8}$	1/(s T)	8.6×10^{-8}
(Ratio)	$(1/2\pi)\gamma_p$	$42.5774813(37)$	MHz/T	8.6×10^{-8}

Table 1.1-8 Properties of the neutron

Quantity	Symbol and relation	Numerical value	Units	Relative standard uncertainty
Neutron mass	m_n	$1.67492728(29) \times 10^{-27}$ $1.00866491560(55)$	kg u	1.7×10^{-7} 5.5×10^{-10}
Energy equivalent	$m_n c^2$	$1.50534957(26) \times 10^{-10}$ $939.565360(81)$	J MeV	1.7×10^{-7} 8.6×10^{-8}
Neutron–electron mass ratio	m_n/m_e	$1838.6836598(13)$		7.0×10^{-10}
Neutron–proton mass ratio	m_n/m_p	$1.00137841870(58)$		5.8×10^{-10}
Molar mass	$M(n) = N_A m_n$	$1.00866491560(55) \times 10^{-3}$	kg	5.5×10^{-10}
Compton wavelength	$\lambda_{C,n} = h/(m_n c)$	$1.3195909067(88) \times 10^{-15}$	m	6.7×10^{-9}
(Ratio)	$(1/2\pi)\lambda_{C,n}$	$0.2100194157(14) \times 10^{-15}$	m	6.7×10^{-9}
Magnetic moment	μ_n	$-0.96623645(24) \times 10^{-26}$	J/T	2.5×10^{-7}
Ratio of magnetic moment to Bohr magneton	μ_n/μ_B	$-1.04187563(25) \times 10^{-3}$		2.4×10^{-7}
Ratio of magnetic moment to nuclear magneton	μ_n/μ_N	$-1.91304273(45)$		2.4×10^{-7}
Ratio of magnetic moment to electron magnetic moment	μ_n/μ_e	$1.04066882(25) \times 10^{-3}$		2.4×10^{-7}
Ratio of magnetic moment to proton magnetic moment	μ_n/μ_p	$-0.68497934(16)$		2.4×10^{-7}
g-factor	$g_n = 2\mu_n/\mu_N$	$-3.82608546(90)$		2.4×10^{-7}
Gyromagnetic ratio	$\gamma_n = 2\lvert\mu_n\rvert/\hbar$	$1.83247183(46) \times 10^{8}$	1/(s T)	2.5×10^{-7}
(Ratio)	$(1/2\pi)\gamma_n$	$29.1646950(73)$	MHz/T	2.5×10^{-7}

Table 1.1-9 Properties of the alpha particle

Quantity	Symbol and relation	Numerical value	Units	Relative standard uncertainty
Alpha particle mass [a]	m_α	$6.6446565(11) \times 10^{-27}$ $4.001506179149(56)$	kg u	1.7×10^{-7} 1.4×10^{-11}
Energy equivalent of alpha particle mass	$m_\alpha c^2$	$5.9719194(10) \times 10^{-10}$ $3727.37917(32)$	J MeV	1.7×10^{-7} 8.6×10^{-8}
Ratio of alpha particle mass to electron mass	m_α / m_e	$7294.2995363(32)$		4.4×10^{-10}
Ratio of alpha particle mass to proton mass	m_α / m_p	$3.97259968907(52)$		1.3×10^{-10}
Alpha particle molar mass	$M(\alpha) = N_A m_\alpha$	$4.001506179149(56) \times 10^{-3}$	kg/mol	1.4×10^{-11}

[a] The mass of the alpha particle in units of the atomic mass unit u is given by $m_\alpha = A_r(\alpha)$ u; in words, the alpha particle mass is given by the relative atomic mass $A_r(\alpha)$ of the alpha particle, multiplied by the atomic mass unit u

References

1.1 P. J. Mohr, B. N. Taylor: CODATA recommended values of the fundamental physical constants, Rev. Mod. Phys. (2004) (in press)

1.2 NIST Physics Laboratory: Web pages of the Fundamental Constants Data Center, http://physics.nist.gov/constants

1.2. The International System of Units (SI), Physical Quantities, and Their Dimensions

In this chapter, we introduce the International System of Units (SI) on the basis of the SI brochure "Le Système international d'unités (SI)" [2.1], supplemented by [2.2]. We give a short review of how the SI was worked out and who is responsible for the further development of the system. Following the above-mentioned publications, we explain the concepts of base physical quantities and derived physical quantities on which the SI is founded, and present a detailed description of the SI base units and of a large selection of SI derived units. We also discuss a number of non-SI units which still are in use, especially in some specialized fields. A table (Table 1.2-17) presenting the values of various energy equivalents closes the section.

1.2.1 **The International System of Units (SI)** ... 11

1.2.2 **Physical Quantities** ... 12

1.2.3 **The SI Base Units** ... 13
- 1.2.3.1 Unit of Length: the Meter ... 13
- 1.2.3.2 Unit of Mass: the Kilogram ... 14
- 1.2.3.3 Unit of Time: the Second ... 14
- 1.2.3.4 Unit of Electric Current: the Ampere ... 14
- 1.2.3.5 Unit of (Thermodynamic) Temperature: the Kelvin ... 14
- 1.2.3.6 Unit of Amount of Substance: the Mole ... 14
- 1.2.3.7 Unit of Luminous Intensity: the Candela ... 15

1.2.4 **The SI Derived Units** ... 16

1.2.5 **Decimal Multiples and Submultiples of SI Units** ... 19

1.2.6 **Units Outside the SI** ... 20
- 1.2.6.1 Units Used with the SI ... 20
- 1.2.6.2 Other Non-SI Units ... 20

1.2.7 **Some Energy Equivalents** ... 24

References ... 25

1.2.1 The International System of Units (SI)

All data in this handbook are given in the International System of Units (Système International d'Unités), abbreviated internationally to SI, which is the modern metric system of measurement and is acknowledged worldwide. The system of SI units was introduced by the General Conference of Weights and Measures (Conférence Générale des Poids et Mesures), abbreviated internationally to CGPM, in 1960. The system not only is used in science, but also is dominant in technology, industrial production, and international commerce and trade.

Who takes care of this system of SI units?

The Bureau International des Poids et Mesures (BIPM), which has its headquarters in Sèvres near Paris, has taken on a commitment to ensure worldwide unification of physical measurements. Its function is thus to:

- establish fundamental standards and scales for the measurement of the principal physical quantities and maintain the international prototypes;
- carry out comparison of national and international standards;
- ensure the coordination of the corresponding measuring techniques;
- carry out and coordinate measurements of the fundamental physical constants relevant to those activities.

The BIPM operates under the exclusive supervision of the Comité International des Poids et Mesures (CIPM), which itself comes under the authority of the

Conférence Générale des Poids et Mesures and reports to it on the work accomplished by the BIPM. The BIPM itself was set up by the Convention du Mètre signed in Paris in 1875 by 17 states during the final session of the Conference on the Meter. The convention was amended in 1921.

Delegates from all member states of the Convention du Mètre attend the Conférence Générale, which, at present, meets every four years. The function of these meetings is to:

- discuss and initiate the arrangements required to ensure the propagation and improvement of the International System of Units
- confirm the results of new fundamental metrological determinations and confirm various scientific resolutions with international scope
- take all major decisions concerning the finance, organization, and development of the BIPM.

The CIPM has 18 members, each from a different state; at present, it meets every year. The officers of this committee present an annual report on the administrative and financial position of the BIPM to the governments of the member states of the Convention du Mètre. The principal task of the CIPM is to ensure worldwide uniformity in units of measurement. It does this by direct action or by submitting proposals to the CGPM.

The BIPM publishes monographs on special metrological subjects and the brochure *Le Système international d'unités (SI)* [2.1,2], which is periodically updated and in which all decisions and recommendations concerning units are collected together.

The scientific work of the BIPM is published in the open scientific literature, and an annual list of publications appears in the *Procès-Verbaux* of the CIPM.

Since 1965, *Metrologica*, an international journal published under the auspices of the CIPM, has printed articles dealing with scientific metrology, improvements in methods of measurements, and work on standards and units, as well as reports concerning the activities, decisions, and recommendations of the various bodies created under the Convention du Mètre.

1.2.2 Physical Quantities

Physical quantities are tools which allow us to specify and quantify the properties of physical objects and to model the events, phenomena, and patterns of behavior of objects in nature and in technology. The system of physical quantities used with the SI units is dealt by Technical Committee 12 of the International Organization for Standardization (ISO/TC 12). Since 1955, ISO/TC 12 has published a series of international standards on quantities and their units, in which the use of SI units is strongly recommended.

How are Physical Quantities Defined?

It turns out that it is possible to divide the system of all known physical quantities into two groups:

- a small number of *base quantities*;
- a much larger number of other quantities, which are called *derived quantities*.

The derived quantities are introduced into physics unambiguously by a defining equation in terms of the base quantities; the relationships between the derived quantities and the base quantities are expressed in a series of equations, which contain a good deal of our knowledge of physics but are used in this system as the defining equations for new physical quantities. One might say that, in this system, physics is described in the rather low-dimensional space of a small number of base quantities.

Base quantities, on the other hand, cannot be introduced by a defining equation; they cannot be traced back to other quantities; this is what we mean by calling them "base". How can base quantities then be introduced unambiguously into physics at all?

Base physical quantities are introduced into physics in three steps:

- We borrow the qualitative meaning of the word for a base quantity from the meaning of the corresponding word in everyday language.
- We specify this meaning by indicating an appropriate method for measuring the quantity. For example, length is measured by a measuring rule, and time is measured by a clock.
- We fix a unit for this quantity, which allows us to communicate the result of a measurement. Length, for example, is measured in meters; time is measured in seconds.

On the basis of these three steps, it is expected that everyone will understand what is meant when the name of a base quantity is mentioned.

In fact, the number of base quantities chosen and the selection of the quantities which are considered as base quantities are a matter of expediency; in different fields and applications of physics, it might well be expedient to use different numbers of base quantities and different selections of base quantities. It should be kept in mind, however, that the number and selection of base quantities are only a matter of different representations of physics; physics itself is not affected by the choice that is made.

Today, many scientists therefore prefer to use the *conventional system* recommended by ISO. This system uses seven base quantities, which are selected according to the seven base units of the SI system. Table 1.2-1 shows the recommended names, symbols, and measuring devices for the conventional seven base quantities.

Table 1.2-1 The ISO recommended base quantities

Name of quantity	Symbol	Measured by
Length	l	A measuring rule
Time	t	A clock
Mass	m	A balance
Electrical current	I	A balance
Temperature	T	A thermometer
Particle number	N	Counting
Luminous intensity	I_v	A photometer

All other physical quantities can then be defined as derived quantities; this means they can be defined by equations in terms of the seven base quantities. Within this "conventional system", the set of all defining equations for the derived physical quantities also defines the units for the derived quantities in terms of the units of the base quantities. This is the great advantage of the "conventional system".

The quantity velocity v, for example, is defined by the equation

$$v = \frac{dl}{dt} .$$

In this way, velocity is traced back to the two base quantities length l and time t. On the right-hand side of this equation, we have a differential of length dl divided by a differential of time dt. The algebraic combination of the base quantities in the defining equation for a derived quantity is called the *dimensions* of the derived quantity. So velocity has the dimensions length/time, acceleration has the dimensions length/time squared, and so on.

Data for a physical quantity are always given as a product of a number (the *numerical value* of the physical quantity) and a unit in which the quantity has been measured.

1.2.3 The SI Base Units

Formal definitions of the seven SI base units have to be approved by the CGPM. The first such definition was approved in 1889. These definitions have been modified, however, from time to time as techniques of measurement have evolved and allowed more accurate realizations of the base units. Table 1.2-2 summarizes the present status of the SI base units and their symbols.

In the following, the current definitions of the base units adopted by the CGPM are shown in detail, together with some explanatory notes. Related decisions which clarify these definitions but are not formally part of them are also shown indented, but in a font of normal weight.

Table 1.2-2 The seven SI base units and their symbols

Base quantity	Symbol for quantity	Unit	Symbol for unit
Length	l	metre	m
Time	t	second	s
Mass	m	kilogram	kg
Electrical current	I	ampere	A
Temperature	T	kelvin	K
Particle number	n	mole	mol
Luminous intensity	I_v	candela	cd

1.2.3.1 Unit of Length: the Meter

The unit of length, the meter, was defined in the first CGPM approval in 1889 by an international prototype: the length of a bar made of a platinum–iridium alloy defined a length of 1 m. In 1960 this definition was replaced by a definition based upon a wavelength of krypton-86 radiation. Since 1983 (17th CGPM), the meter has been defined as:

> *The metre is the length of the path travelled by light in vacuum during a time interval of 1/299 792 458 of a second.*

As a result of this definition, the fundamental constant "speed of light in vacuum c" is fixed at exactly 299 792 458 m/s.

1.2.3.2 Unit of Mass: the Kilogram

Since the first CGPM in 1889, the unit of mass, the kilogram, has been defined by an international prototype, a metal block made of a platinum–iridium alloy, kept at the BIPM at Sèvres. The relevant declaration was modified slightly at the third CGPM in 1901 to confirm that:

The kilogram is the unit of mass; it is equal to the mass of the international prototype of the kilogram.

1.2.3.3 Unit of Time: the Second

The unit of time, the second, was originally considered to be the fraction 1/86 400 of the mean solar day. Measurements, however, showed that irregularities in the rotation of the Earth could not be taken into account by theory, and these irregularities have the effect that this definition does not allow the required accuracy to be achieved. The same turned out to be true for other definitions based on astronomical data. Experimental work, however, had already shown that an atomic standard of time interval, based on a transition between two energy levels of an atom or a molecule, could be realized and reproduced much more precisely. Therefore, the 13th CGPM (1967 – 1968) replaced the definition of the second by:

The second is the duration of 9 192 631 770 periods of the radiation corresponding to the transition between the two hyperfine levels of the ground state of the caesium 133 atom.

At its 1997 meeting, the CIPM affirmed that:

This definition refers to a caesium atom at rest at a temperature of 0 K.

This note was intended to make it clear that the definition of the SI second is based on a Cs atom unperturbed by black-body radiation, that is, in an environment whose temperature is 0 K.

1.2.3.4 Unit of Electric Current: the Ampere

"International" electrical units for current and resistance were introduced by the International Electrical Congress in Chicago as early as in 1893 and were confirmed by an international conference in London in 1908. They were replaced by an "absolute" definition of the ampere as the unit for electric current at the 9th CGPM in 1948, which stated:

The ampere is that constant current which, if maintained in two straight parallel conductors of infinite length, of negligible circular cross-section, and placed 1 meter apart in vacuum, would produce between these conductors a force equal to 2×10^{-7} newton per meter of length.

As a result of this definition, the fundamental constant "magnetic field constant μ_0" (also known as the permeability of free space) is fixed at exactly $4\pi \times 10^{-7}\,\mathrm{N/A^2}$.

1.2.3.5 Unit of (Thermodynamic) Temperature: the Kelvin

The definition of the unit of (thermodynamic) temperature was given in substance by the 10th CGPM in 1954, which selected the triple point of water as the fundamental fixed point and assigned to it the temperature 273.16 K, so defining the unit. After smaller amendments, made at the 13th CGPM in 1967–1968, the definition of the unit of (thermodynamic) temperature reads

The kelvin, unit of thermodynamic temperature, is the fraction 1/273.16 of the thermodynamic temperature of the triple point of water.

Because of the way temperature scales used to be defined, it remains common practice to express a thermodynamic temperature, symbol T, in terms of its difference from the reference temperature $T_0 = 273.15\,\mathrm{K}$, the ice point. This temperature difference is called the Celsius temperature, symbol t, and is defined by the equation $t = T - T_0$. The unit of Celsius temperature is the degree Celsius, symbol °C, which is, by definition, equal in magnitude to the kelvin. A difference or interval of temperature may therefore be expressed either in kelvin or in degrees Celsius.

The numerical value of a Celsius temperature t expressed in degrees Celsius is given by $t\,(°\mathrm{C}) = T\,(\mathrm{K}) - 273.15$.

1.2.3.6 Unit of Amount of Substance: the Mole

The "amount of substance" of a sample is understood as a measure of the number of elementary entities (for example atoms or molecules) that the sample consists of. Owing to the fact that on macroscopic scales this number cannot be counted directly in most cases, one has to relate this quantity "amount of substance" to a more easily measurable quantity, the mass of a sample of that substance.

On the basis of an agreement between the International Union of Pure and Applied Physics (IUPAP) and the International Union of Pure and Applied Chemistry (IUPAC) in 1959/1960, physicists and chemists have ever since agreed to assign, by definition, the value 12, exactly, to the relative atomic mass (formerly called "atomic weight") of the isotope of carbon with mass number 12 (carbon-12, ^{12}C). The scale of the masses of all other atoms and isotopes based on this agreement has been called, since then, the scale of relative atomic masses.

It remains to define the unit of the "amount of substance" in terms of the mass of the corresponding amount of the substance. This is done by fixing the mass of a particular amount of carbon-12; by international agreement, this mass has been fixed at 0.012 kg. The corresponding unit of the quantity "amount of substance" has been given the name "mole" (symbol mol).

On the basis of proposals by IUPAC, IUPAP, and ISO, the CIPM formulated a definition of the mole in 1967 and confirmed it in 1969. This definition was adopted by the 14th CGPM in 1971 in two statements:

1. *The mole is the amount of substance of a system which contains as many elementary entities as there are atoms in* 0.012 *kilogram of carbon-12; its symbol is "mol".*
2. *When the mole is used, the elementary entities must be specified and may be atoms, molecules, ions, electrons, other particles, or specified groups of such particles.*

In 1980, the CIPM approved the report of the Comité Consultatif des Unités (CCU), which specified that:

> In this definition, it is understood that unbound atoms of carbon-12, at rest and in their ground state, are referred to.

1.2.3.7 Unit of Luminous Intensity: the Candela

The base unit candela allows one to establish a quantitative relation between radiometric and photometric measurements of light intensities. In physics and chemistry, the intensities of radiation fields of various natures are normally determined by radiometry; in visual optics, in lighting engineering, and in the physiology of the visual system, however, it is necessary to assess the intensity of the radiation field by photometric means.

There are, in fact, three different ways to quantify the intensity of a radiation beam. One way is to measure the "radiant intensity" I_e, defined as the radiant flux $\Delta\Phi_e$ per unit solid angle $\Delta\Omega$ of the beam. The subscript "e" stands for "energetic". Here, the radiant flux Φ_e is defined as the energy of the radiation per unit time, and is accordingly measured in units of watts (W). The radiant intensity I_e, therefore, has the dimensions of energy per time per solid angle, and is measured in the derived unit "watt per steradian" (W/sr).

Another way to quantify the intensity of a beam of radiation is to measure the "particle intensity" I_p, which is defined as the particle flux Φ_p divided by the solid angle $\Delta\Omega$ of the beam. The subscript "p" stands for "particle". The particle flux Φ_p itself is measured by counting the number of particles per unit time in the beam; in the case of a light beam, for example, the particles are photons. The corresponding SI unit for the particle flux is seconds^{-1} (1/s). The quantity particle intensity I_p therefore has the dimensions of number per time per solid angle; the corresponding derived SI unit for the particle intensity I_p is "seconds^{-1} times steradian^{-1}" (1/(s sr)).

In addition to these two radiometric assessments of the beam intensity, for beams of visible light there is a third possibility, which is to quantify the intensity of the beam by the intensity of visual perception by the human eye. Physical quantities connected with this physiological type of assessment are called photometric quantities, in contrast to the two radiometric quantities described above. In photometry, the intensity of the beam is called the "luminous intensity" I_v. The subscript "v" stands for "visual". The luminous intensity I_v is an ISO recommended base quantity; the corresponding SI base unit is the candela (cd). The luminous flux Φ_v is determined as the product of the luminous intensity and the solid angle. Its dimensions therefore are luminous intensity times solid angle, so that the SI unit of the luminous flux Φ_v turns out to be "candela times steradian" (cd sr). A derived unit, the lumen (lm), such that 1 lm = 1 cd sr, has been introduced for this product.

Table 1.2-3 summarizes the names, definitions, and SI units for the most frequently used radiometric and photometric quantities in radiation physics.

The history of the base unit candela is as follows. Before 1948, the units for photometric measurements were be based on flame or incandescent-filament standards. They were replaced initially by the "new candle" based on the luminance of a Planckian radiator (a blackbody radiator) at the temperature of freezing platinum. This modification was ratified in 1948 by the 9th CGPM, which also adopted the new international name for the base unit of luminous intensity, the candela, and its symbol cd. The 13th CGPM gave an amended version of the 1948 definition in 1967.

Table 1.2-3 Radiometric and photometric quantities in radiation physics

Quantity	Symbol and definition	Dimensions	SI unit	Symbol for unit
Radiant flux	$\Phi_e = \Delta E/\Delta t$ [a]	Power = energy/time	watt	$W = J/s$
Particle flux, activity	$\Phi_p = \Delta N_p/\Delta t$	1/time	$second^{-1}$	$1/s$
Luminous flux	$\Phi_v = I_v \Omega$ [b]	Luminous intensity times solid angle	lumen	$lm = cd\,sr$
Radiant intensity	$I_e = \Delta\Phi_e/\Delta\Omega$	Power/solid angle	watt/steradian	W/sr
Particle intensity	$I_p = \Delta\Phi_p/\Delta\Omega$	$(Time\ times\ solid\ angle)^{-1}$	$(second\ times\ steradian)^{-1}$	$1/(s\,sr)$
Luminous intensity	I_v, basic quantity	Luminous intensity	candela	cd
Radiance [c]	$L_e = \Delta I_e(\varphi)/[\Delta A_1 g(\varphi)]$ [d]	Power per source area and solid angle	watt/(meter^2 times steradian)	$W/(m^2\,sr) = kg/(s^3\,sr)$
Particle radiance [c]	$L_p = \Delta I_p(\varphi)/[\Delta A_1 g(\varphi)]$ [d]	(Time times area times solid angle)$^{-1}$	1/(second times meter^2 times steradian)	$1/(s\,m^2\,sr)$
Luminance [c]	$L_v = \Delta I_v(\varphi)/[\Delta A_1 g(\varphi)]$ [d]	Luminous intensity/source area	candela/meter^2	cd/m^2
Irradiance	$E_e = \Delta\Phi_e/\Delta A_2$ [e]	Power/area	watt/meter^2	W/m^2
Particle irradiance	$E_p = \Delta\Phi_p/\Delta A_2$ [e]	Number of particles per (time times area)	1/(second times meter^2)	$1/(s\,m^2)$
Illuminance	$E_v = \Delta\Phi_v/\Delta A_2$ [e]	Luminous flux per area	lux = lumen/meter^2	$lx = lm/m^2 = cd\,sr/m^2$

[a] The symbol E stands for the radiant energy (see Table 1.2-5).

[b] I_v stands for the luminous intensity, and Ω stands for the solid angle (see Table 1.2-5).

[c] The radiance L_e, particle radiance L_p, and luminance L_v are important characteristic properties of sources, not radiation fields. For a black-body source, the radiance L_e, for example, is dependent only on the frequency of the radiation and the temperature of the black body. The dependence is given by Planck's radiation law. In optical imaging, the radiance L_e of an object turns out to show an invariant property. In correct imaging, the image always radiates with the same radiance L_e as the object, independent of the magnification.

[d] φ is the angle between the direction of the beam axis and the direction perpendicular to the source area; A_1 indicates the area of the source; and $g(\varphi)$ is the directional characteristic of the source.

[e] A_2 indicates the irradiated area or the area of the detector.

Because of experimental difficulties in realizing a Planckian radiator at high temperatures and because of new possibilities in the measurement of optical radiation power, the 16th CGPM in 1979 adopted a new definition of the candela as follows:

The candela is the luminous intensity, in a given direction, of a source that emits monochromatic radiation of frequency 540×10^{12} *hertz and that has a radiant intensity in that direction of* $1/683$ *watt per steradian.*

1.2.4 The SI Derived Units

The SI derived units are the SI units for derived physical quantities. In accordance with the defining equations for derived physical quantities in terms of the base physical quantities, the units for derived quantities can be expressed as products or ratios of the units for the base quantities. Table 1.2-4 shows some examples of SI derived units in terms of SI base units.

For convenience, certain derived units, which are listed in Table 1.2-5, have been given special names and symbols. Among these, the last four entries in Table 1.2-5 are of particular note, since they were accepted by the 15th (1975), 16th (1979), and 21st (1999) CGPMs specifically with a view to safeguarding human health.

In Tables 1.2-5 and 1.2-6, the final column shows how the SI units concerned may be expressed in terms of SI base units. In this column, factors such as m^0 and kg^0, etc., which are equal to 1, are not shown explicitly.

The special names and symbols for derived units listed in Table 1.2-5 may themselves be used to express other derived units: Table 1.2-6 shows some examples. The special names and symbols provide a compact form for the expression of units which are used frequently.

Table 1.2-4 Examples of SI derived units (for derived physical quantities) in terms of base units

Derived quantity	Defining equation	Name of SI derived unit	Symbol for unit
Area	$A = l_1 l_2$	square meter	m^2
Volume	$V = l_1 l_2 l_3$	cubic meter	m^3
Velocity	$v = dl/dt$	meter per second	m/s
Acceleration	$a = d^2l/dt^2$	meter per second squared	m/s^2
Angular momentum	$L = \Theta\omega$	meter squared kilogram/second	m^2 kg/s
Wavenumber	$k = 2\pi/\lambda$	reciprocal meter	1/m
Density	$\varrho = m/V$	kilogram per cubic meter	kg/m^3
Concentration (of amount of substance)	Concentration = amount/V	mole per cubic meter	mol/m^3
Current density	$j = I/A$	ampere per square meter	A/m^2
Magnetic exciting field	$H = I/l$	ampere per meter	A/m
Radiance (of a radiation source)	$L_e = \Delta I_e(\varphi)/[\Delta A_1 g(\varphi)]$ [a]	watt per (square meter × steradian)	$W/(m^2\,sr) = kg/(s^3\,sr)$
Luminance (of a light source)	$L_v = \Delta I_v(\varphi)/[\Delta A_1 g(\varphi)]$ [b]	candela per square meter	cd/m^2
Refractive index	$n = c_{mat}/c$	(number one)	1

[a] φ is the angle between the direction of the beam axis and the direction perpendicular to the source area; $I_e(\varphi)$ is the radiant intensity emitted in the direction φ; A_1 is the radiating area of the source; and $g(\varphi)$ is the directional characteristic of the source.

[b] φ is the angle between the direction of the beam axis and the direction perpendicular to the source area; $I_v(\varphi)$ is the luminous intensity emitted in the direction φ; A_1 is the radiating area of the light source; and $g(\varphi)$ is the directional characteristic of the source.

Table 1.2-5 SI derived units with special names and symbols

Derived quantity		SI derived unit			
Name	Symbol	Name	Symbol	Expressed in terms of other SI units	Expressed in terms of SI base units
Plane angle	$\alpha, \Delta\alpha$	radian [a]	rad		m/m = 1 [b]
Solid angle	$\Omega, \Delta\Omega$	steradian [a]	sr [c]		$m^2/m^2 = 1$ [b]
Frequency	ν	hertz	Hz		1/s
Force	F	newton	N		$m\,kg/s^2$
Pressure, stress	P	pascal	Pa	N/m^2	$(1/m)\,kg/s^2$
Energy, work, quantity of heat	E, A, Q	joule	J	N m	$m^2\,kg/s^2$
Power, radiant flux	P, Φ_e	watt	W	J/s	$m^2\,kg/s^3$
Electric charge, quantity of electricity	q, e	coulomb	C		A s
Electric potential difference, electromotive force	V	volt	V	W/A	$(1/A)\,m^2\,kg/s^3$
Capacitance	C	farad	F	C/V	$A^2\,(1/(m^2\,kg))\,s^4$
Electrical resistance	R	ohm	Ω	V/A	$(1/A^2)\,m^2\,kg/s^3$
Electrical conductance	$1/R$	siemens	S	A/V	$A^2\,(1/(m^2\,kg))s^3$
Magnetic flux	Φ	weber	Wb	V s	$(1/A)\,m^2\,kg(1/s^2)$
Magnetic field strength	B	tesla	T	Wb/m^2	$(1/A)\,kg/s^2$
Inductance	L	henry	H	Wb/A	$(1/A^2)\,m^2\,kg/s^2$
Celsius temperature	t	degree Celsius	°C		K; $T/K = t/°C + 273.15$
Luminous flux	Φ_v	lumen	lm	cd sr [c]	(m^2/m^2) cd = cd

Table 1.2-5 SI derived units with special names and symbols, cont.

Derived quantity		SI derived unit			
Name	**Symbol**	**Name**	**Symbol**	**Expressed in terms of other SI units**	**Expressed in terms of SI base units**
Illuminance	$E_v = \Delta\Phi_v/\Delta A$	lux	lx	lm/m^2	(m^2/m^4) cd = cd/m^2
Activity (referred to a radionuclide)	A	becquerel	Bq		1/s
Absorbed dose	D	gray	Gy	J/kg	m^2/s^2
Dose equivalent	H	sievert	Sv	J/kg	m^2/s^2
Catalytic activity		katal	kat		(1/s) mol

[a] The units radian and steradian may be used with advantage in expressions for derived units to distinguish between quantities of different nature but the same dimensions. Some examples of their use in forming derived units are given in Tables 1.2-5 and 1.2-6.

[b] In practice, the symbols rad and sr are used where appropriate, but the derived unit "1" is generally omitted in combination with a numerical value.

[c] In photometry, the name steradian and the symbol sr are frequently retained in expressions for units.

Table 1.2-6 Examples of SI derived units whose names and symbols include SI derived units with special names and symbols

Derived quantity		SI derived unit		
Name	**Symbol**	**Name**	**Symbol**	**Expressed in terms of SI base units**
Dynamic viscosity	η	pascal second	Pa s	(1/m) kg/s
Moment of force	M	newton meter	N m	m^2 kg/s^2
Surface tension	σ	newton per meter	N/m	kg/s^2
Angular velocity	ω	radian per second	rad/s	m/(m s) = 1/s
Angular acceleration	$d\omega/dt$	radian per second squared	rad/s^2	m/(m s^2) = 1/s^2
Heat flux density	q_{th}	watt per square meter	W/m^2	kg/s^3
Heat capacity, entropy	C, S	joule per kelvin	J/K	m^2 kg/(s^2 K)
Specific heat capacity, specific entropy	C_{mass}, S_{mass}	joule per (kilogram kelvin)	J/(kg K)	m^2/(s^2 K)
Specific energy		joule per kilogram	J/kg	m^2/s^2
Energy density	w	joule per cubic meter	J/m^3	(1/m) kg/s^2
Thermal conductivity	λ	watt per (meter kelvin)	W/(m K)	m kg/(s^3 K)
Electric charge density	ρ	coulomb per cubic meter	C/m^3	(1/m^3) s A
Electric field strength	E	volt per meter	V/m	m kg/(s^3 A)
Exciting electric field[b]	D	coulomb per square meter	C/m^2	(1/m^2) s A
Molar energy	E_{mol}	joule per mole	J/mol	m^2 kg/(s^2 mol)
Molar heat capacity, molar entropy	C_{mol}, S_{mol}	joule per (mole kelvin)	J/(mol K)	m^2 kg/(s^2 K mol)
Exposure (X- and γ-rays)		coulomb per kilogram	C/kg	(1/kg) s A
Absorbed dose rate	dD/dt	gray per second	Gy/s	m^2/s^3
Radiant intensity	I_ε	watt per steradian	W/sr	(m^4/m^2) kg/s^3 = m^2 kg/s^3
Radiance[a]	L_e	watt per (square meter steradian)	W/(m^2 sr)	(m^2/m^2) kg/s^3 = kg/s^3
Catalytic (activity) concentration		katal per cubic meter	kat/m^3	(1/(m^3 s)) mol

[a] The radiance is a property of the source of the radiation, not of the radiation field (see footnote c to Table 1.2-3).

[b] also called "electric flux density"

A derived unit can often be expressed in several different ways by combining the names of base units with special names for derived units. This, however, is an algebraic freedom whose use should be limited by common-sense physical considerations. The joule, for example, may formally be written "newton meter" or even "kilogram meter squared per second squared", but in a given situation some forms may be more helpful than others.

In practice, with certain quantities, preference is given to the use of certain special unit names or combinations of unit names, in order to facilitate making a distinction between different quantities that have the same dimensions. For example, the SI unit of frequency is called the hertz rather than the reciprocal second, and the SI unit of angular velocity is called the radian per second rather than the reciprocal second (in this case, retaining the word "radian" emphasizes that the angular velocity is equal to 2π times the rotational frequency). Similarly, the SI unit of moment of force is called the newton meter rather than the joule.

In the field of ionizing radiation, the SI unit of activity is called the becquerel rather than the reciprocal second, and the SI units of absorbed dose and dose equivalent are called the gray and the sievert, respectively, rather than the joule per kilogram. In the field of catalysis, the SI unit of catalytic activity is called the katal rather than the mole per second. The special names becquerel, gray, sievert, and katal were specifically introduced because of the dangers to human health which might arise from mistakes involving the units reciprocal second, joule per kilogram, and mole per second.

1.2.5 Decimal Multiples and Submultiples of SI Units

The 11th CGPM adopted, in 1960, a series of prefixes and prefix symbols for forming the names and symbols of the decimal multiples and submultiples of SI units ranging from 10^{12} to 10^{-12}. Prefixes for 10^{-15} and 10^{-18} were added by the 12th CGPM in 1964, and for 10^{15} and 10^{18} by the 15th CGPM in 1975. The 19th CGPM extended the scale in 1991 from 10^{-24} to 10^{24}. Table 1.2-7 lists all approved prefixes and symbols.

Table 1.2-7 SI prefixes and their symbols

Factor	Name	Symbol
10^{24}	yotta	Y
10^{21}	zeta	Z
10^{18}	exa	E
10^{15}	peta	P
10^{12}	tera	T
10^{9}	giga	G
10^{6}	mega	M
10^{3}	kilo	k
10^{2}	hecto	h
10^{1}	deca	da
10^{-1}	deci	d
10^{-2}	centi	c
10^{-3}	milli	m
10^{-6}	micro	μ
10^{-9}	nano	n
10^{-12}	pico	p
10^{-15}	femto	f
10^{-18}	atto	a
10^{-21}	zepto	z
10^{-24}	yocto	y

1.2.6 Units Outside the SI

The SI base units and SI derived units, including those with special names, have the important advantage of forming a coherent set, with the effect that unit conversions are not required when one is inserting particular values for quantities into equations involving quantities.

Nonetheless, it is recognized that some non-SI units still appear widely in the scientific, technical, and commercial literature, and some will probably continue to be used for many years. Other non-SI units, such as the units of time, are so widely used in everyday life and are so deeply embedded in the history and culture of human beings that they will continue to be used for the foreseeable future. For these reasons, some of the more important non-SI units are listed.

1.2.6.1 Units Used with the SI

In 1996 the CIPM agreed upon a categorization of the units used with the SI into three groups: units accepted for use with the SI, units accepted for use with the SI whose values are obtained experimentally, and other units currently accepted for use with the SI to satisfy the needs of special interests. The three groups are listed in Tables 1.2-8 – 1.2-10.

Table 1.2-9 lists three non-SI units accepted for use with the SI, whose values expressed in SI units must be obtained by experiment and are therefore not known exactly. Their values are given with their combined standard uncertainties, which apply to the last two digits, shown in parentheses. These units are in common use in certain specialized fields.

Table 1.2-10 lists some other non-SI units which are currently accepted for use with the SI to satisfy the needs of commercial, legal, and specialized scientific interests. These units should be defined in relation to SI units in every document in which they are used. Their use is not encouraged.

1.2.6.2 Other Non-SI Units

Certain other non-SI units are still occasionally used. Some are important for the interpretation of older scientific texts. These are listed in Tables 1.2-11 – 1.2-16, but their use is not encouraged.

Table 1.2-8 Non-SI units accepted for use with the International System

Name	Symbol	Value in SI units
minute	min	$1\ \text{min} = 60\ \text{s}$
hour	h	$1\ \text{h} = 60\ \text{min} = 3600\ \text{s}$
day	d	$1\ \text{d} = 24\ \text{h} = 86\,400\ \text{s}$
degree [a]	°	$1° = (\pi/180)\ \text{rad}$
minute of arc	′	$1' = (1/60)° = (\pi/10\,800)\ \text{rad}$
second of arc	″	$1'' = (1/60)' = (\pi/648\,000)\ \text{rad}$
litre [b]	l, L	$1\ \text{l} = 1\ \text{dm}^3 = 10^{-3}\ \text{m}^3$
tonne [c]	t	$1\ \text{t} = 10^3\ \text{kg}$

[a] ISO 31 recommends that the degree be subdivided decimally rather than using the minute and second.

[b] Unfortunately, printers from all over the world seem not to be willing to admit that in some texts it would be very helpful to have distinguishable symbols for "the number 1" and "the letter l". Giving up any further discussion, the 16th CGPM therefore decided in 1979 that the symbol L should also be adopted to indicate the unit litre in order to avoid the risk of confusion between "the number 1" and "the letter l".

[c] This unit is also called the "metric ton" in some countries.

Table 1.2-9 Non-SI units accepted for use with the International System, whose values in SI units are obtained experimentally

Unit	Definition	Symbol	Value in SI units
Electron volt [a]	b	eV	$1\ \text{eV} = 1.60217653(14) \times 10^{-19}\ \text{J}$
Unified atomic mass unit [a]	c	u	$1\ \text{u} = 1.66053886(28) \times 10^{-27}\ \text{kg}$
Astronomical unit [d]	e	ua	$1\ \text{ua} = 1.49597870691(30) \times 10^{11}\ \text{m}$

[a] For the electron volt and the unified atomic mass unit, the values are quoted from the CODATA recommended values 2002 (see Chapt. 1.1).

[b] The electron volt is the kinetic energy acquired by an electron in passing through a potential difference of 1 V in vacuum.

[c] The unified atomic mass unit is equal to 1/12 of the mass of an unbound atom of the nuclide ^{12}C, at rest and in its ground state. In the field of biochemistry, the unified atomic mass unit is also called the dalton, symbol Da.

[d] The value given for the astronomical unit is quoted from the IERS Convention (1996).

[e] The astronomical unit is a unit of length approximately equal to the mean Earth–Sun distance. Its value is such that, when it is used to describe the motion of bodies in the solar system, the heliocentric gravitational constant is $(0.01720209895)^2\ \text{ua}^3/\text{d}^2$.

Table 1.2-10 Other non-SI units currently accepted for use with the International System

Unit	Symbol	Value in SI units
nautical mile[a]		1 nautical mile = 1852 m
knot		1 knot = 1 nautical mile per hour = (1852/3600) m/s
are[b]	a	$1\,a = 1\,dam^2 = 10^2\,m^2$
hectare[b]	ha	$1\,ha = 1\,hm^2 = 10^4\,m^2$
Bar[c]	bar	1 bar = 0.1 MPa = 100 kPa = 1000 hPa = 10^5 Pa
angstrom	Å	$1\,Å = 0.1\,nm = 10^{-10}\,m$
barn[d]	b	$1\,b = 100\,fm^2 = 10^{-28}\,m^2$

[a] The nautical mile is a special unit employed for marine and aerial navigation to express distance. The conventional value given above was adopted by the First International Extraordinary Hydrographic Conference, Monaco, 1929, under the name "international nautical mile". As yet there is no internationally agreed symbol. This unit was originally chosen because one nautical mile on the surface of the Earth subtends approximately one minute of arc at the center.

[b] The units are and hectare and their symbols were adopted by the CIPM in 1879 and are used to express areas of land.

[c] The bar and its symbol were included in Resolution 7 of the 9th CGPM (1948).

[d] The barn is a special unit employed in nuclear physics to express effective cross sections.

Table 1.2-11 Derived CGS units with special names

Unit	Symbol	Value in SI units
erg[a]	erg	$1\,erg = 10^{-7}\,J$
dyne[a]	dyn	$1\,dyn = 10^{-5}\,N$
poise[a]	P	$1\,P = 1\,dyn\,s/cm^2 = 0.1\,Pa\,s$
stokes	St	$1\,St = 1\,cm^2/s = 10^{-4}\,m^2/s$
gauss[b]	G	$1\,G \equiv 10^{-4}\,T$
oersted[b]	Oe	$1\,Oe \equiv (1000/4\pi)\,A/m$
maxwell[b]	Mx	$1\,Mx \equiv 10^{-8}\,Wb$
stilb[a]	sb	$1\,sb = 1\,cd/cm^2 = 10^4\,cd/m^2$
phot	ph	$1\,ph = 10^4\,lx$
gal[c]	Gal	$1\,Gal = 1\,cm/s^2 = 10^{-2}\,m/s^2$

[a] This unit and its symbol were included in Resolution 7 of the 9th CGPM (1948).

[b] This unit is part of the "electromagnetic" three-dimensional CGS system and cannot strictly be compared with the corresponding unit of the International System, which has four dimensions if only mechanical and electrical quantities are considered. For this reason, this unit is linked to the SI unit using the mathematical symbol for "equivalent to" (≡) here.

[c] The gal is a special unit employed in geodesy and geophysics to express the acceleration due to gravity.

Table 1.2-11 deals with the relationship between CGS units and SI units, and lists those CGS units that were assigned special names. In the field of mechanics, the CGS system of units was built upon three quantities and their corresponding base units: the centimeter, the gram, and the second. In the field of electricity and magnetism, units were expressed in terms of these three base units. Because this can be done in different ways, this led to the establishment of several different systems, for example the CGS electrostatic system, the CGS electromagnetic system, and the CGS Gaussian system. In those three systems, the system of quantities used and the corresponding system of defining equations for the derived quantities differ from those used with SI units.

Table 1.2-12 deals with the *natural units*, which are based directly on fundamental constants or combinations of fundamental constants. Like the CGS system, this system is based on mechanical quantities only. The numerical values in SI units are given here according to the 2002 CODATA adjustment.

Table 1.2-13 presents numerical values in SI units for some of the most frequently used *atomic units* (a.u.), again based on the 2002 CODATA adjustment.

Table 1.2-14 presents numerical values in SI units (based on the 2002 CODATA adjustment) for some X-ray-related quantities used in crystallography.

Table 1.2-12 Natural units (n.u.)

Unit	Symbol and definition	Value in SI units
n.u. of velocity: speed of light in vacuum	c	299 792 458 m/s
n.u. of action: reduced Planck constant	$\hbar = h/2\pi$	$1.05457168(18) \times 10^{-34}\,J\,s$ $6.58211915(56) \times 10^{-16}\,eV\,s$
n.u. of mass: electron mass	m_e	$9.1093826(16) \times 10^{-31}\,kg$
n.u. of energy	$m_e c^2$	$8.1871047(14) \times 10^{-14}\,J$ 0.510998918(44) MeV
n.u. of momentum	$m_e c$	$2.73092419(47) \times 10^{-22}\,kg\,m/s$ 0.510998918(44) MeV/c
n.u. of length	$\lambda\!\!\!^{-}_C = \hbar/m_e c$	$386.1592678(26) \times 10^{-15}\,m$
n.u. of time	$\hbar/m_e c^2$	$1.2880886677(86) \times 10^{-21}\,s$

Table 1.2-13 Atomic units (a.u.)

Unit	Symbol and definition	Value in SI units
a.u. of charge: elementary charge	e	$1.60217653(14) \times 10^{-19}$ C
a.u. of mass: electron mass	m_e	$9.1093826(16) \times 10^{-31}$ kg
a.u. of action: reduced Planck constant	$\hbar = h/2\pi$	$1.05457168(18) \times 10^{-34}$ J s
a.u. of length, 1 bohr: Bohr radius	$a_0 = \alpha/(4\pi R_\infty)$	$0.5291772108(18) \times 10^{-10}$ m
a.u. of energy, 1 hartree: Hartree energy[a]	E_H	$4.35974417(75) \times 10^{-18}$ J
a.u. of time	$\hbar/E_H$	$2.418884326505(16) \times 10^{-17}$ s
a.u. of force	E_H/a_0	$8.2387225(14) \times 10^{-8}$ N
a.u. of velocity	$\alpha c = a_0 E_H/\hbar$	$2.1876912633(73) \times 10^{6}$ m/s
a.u. of momentum	$\hbar/a_0$	$1.99285166(34) \times 10^{-24}$ kg m/s
a.u. of current	$eE_H/\hbar$	$6.62361782(57) \times 10^{-3}$ A
a.u. of charge density	e/a_0^3	$1.081202317(93) \times 10^{12}$ C/m^3
a.u. of electric potential	E_H/e	27.2113845(23) V
a.u. of electric field	$E_H/(ea_0)$	$5.14220642(44) \times 10^{11}$ V/m
a.u. of electric dipole moment	ea_0	$8.47835309(73) \times 10^{-30}$ C m
a.u. of electric polarizability	$e^2a_0^2/E_H$	$1.648777274(16) \times 10^{-41}$ C^2 m^2/J
a.u. of magnetic field B	$\hbar/(ea_0^2)$	$2.35051742(20) \times 10^{5}$ T
a.u. of magnetic dipole moment ($2\mu_B$)	$2\mu_B = \hbar e/m_e$	$1.85480190(16) \times 10^{-23}$ J/T
a.u. of magnetizability	$e^2a_0^2/m_e$	$7.89103660(13) \times 10^{-29}$ J/T^2
a.u. of permittivity	$e^2/(a_0 E_H)$	Fixed by definition as: $10^7/c^2 = 1.112650056\ldots10^{-10}$ F/m

[a] The Hartree energy is defined as $E_H = e^2/(4\pi\varepsilon_0 a_0) = 2R_\infty hc = \alpha^2 m_e c^2$.

Table 1.2-14 Units of some special X-ray-related quantities

Unit	Definition	Symbol	Value in SI units
Cu X unit	$\lambda(\mathrm{CuK}\alpha_1)/1537.400$	$\mathrm{xu}(\mathrm{CuK}\alpha_1)$	$1.00207710(29) \times 10^{-13}$ m
Mo X unit	$\lambda(\mathrm{MoK}\alpha_1)/707.831$	$\mathrm{xu}(\mathrm{MoK}\alpha_1)$	$1.00209966(53) \times 10^{-13}$ m
angstrom star	$\lambda(\mathrm{WK}\alpha_1)/0.2090100$	Å*	$1.00001509(90) \times 10^{-10}$ m
Lattice parameter[a] of Si (in vacuum, at 22.5 °C)		a	$543.102122(20) \times 10^{-12}$ m
(220) lattice spacing of Si (in vacuum, at 22.5 °C)	$d_{220} = a/\sqrt{8}$	d_{220}	$192.0155965(70) \times 10^{-12}$ m
Molar volume of Si (in vacuum, at 22.5 °C)	$V_m(\mathrm{Si}) = N_A a^3/8$	$V_m(\mathrm{Si})$	$12.0588382(24) \times 10^{-6}$ m^3/mol

[a] This is the lattice parameter (unit cell edge length) of an ideal single crystal of naturally occurring silicon free from impurities and imperfections, and is deduced from measurements on extremely pure, nearly perfect single crystals of Si by correcting for the effects of impurities.

Table 1.2-15 lists some other units which are common in older texts. For current texts, it should be noted that if these units are used, the advantages of the SI are lost. The relation of these units to SI units should be specified in every document in which they are used.

For some selected quantities, there exists an international agreement that the numerical values of these quantities measured in SI units are fixed at the values given in Table 1.2-16.

Table 1.2-15 Examples of other non-SI units

Unit	Symbol	Value in SI units
curie [a]	Ci	$1\,\mathrm{Ci} = 3.7\times10^{10}\,\mathrm{Bq}$
röntgen [b]	R	$1\,\mathrm{R} = 2.58\times10^{-4}\,\mathrm{C/kg}$
rad [c,d]	rad	$1\,\mathrm{rad} = 1\,\mathrm{cGy} = 10^{-2}\,\mathrm{Gy}$
rem [d,e]	rem	$1\,\mathrm{rem} = 1\,\mathrm{cSv} = 10^{-2}\,\mathrm{Sv}$
X unit [f]		$1\,\mathrm{X\,unit} \cong 1.002\times10^{-4}\,\mathrm{nm}$
gamma [d]	γ	$1\,\gamma = 1\,\mathrm{nT} = 10^{-9}\,\mathrm{T}$
jansky	Jy	$1\,\mathrm{Jy} = 10^{-26}\,\mathrm{W/(m^2\,Hz)}$
fermi [d]		$1\,\mathrm{fermi} = 1\,\mathrm{fm} = 10^{-15}\,\mathrm{m}$
metric carat [g]		$1\,\mathrm{metric\ carat} = 200\,\mathrm{mg} = 2\times10^{-4}\,\mathrm{kg}$
torr	Torr	$1\,\mathrm{Torr} = (101\,325/760)\,\mathrm{Pa}$
standard atmosphere	atm [h]	$1\,\mathrm{atm} = 101\,325\,\mathrm{Pa}$
calorie	cal	[i]
micron [j]	μ	$1\,\mu = 1\,\mu\mathrm{m} = 10^{-6}\,\mathrm{m}$

[a] The curie is a special unit employed in nuclear physics to express the activity of radionuclides.

[b] The röntgen is a special unit employed to express exposure to X-ray or γ radiation.

[c] The rad is a special unit employed to express absorbed dose of ionizing radiation. When there is a risk of confusion with the symbol for the radian, rd may be used as the symbol for 10^{-2} Gy.

[e] The rem is a special unit used in radioprotection to express dose equivalent.

[f] The X unit was employed to express wavelengths of X-rays. Its relationship to SI units is an approximate one.

[d] Note that this non-SI unit is exactly equivalent to an SI unit with an appropriate submultiple prefix.

[g] The metric carat was adopted by the 4th CGPM in 1907 for commercial dealings in diamonds, pearls, and precious stones.

[h] Resolution 4 of the 10th CGPM, 1954. The designation "standard atmosphere" for a reference pressure of 101 325 Pa is still acceptable.

[i] Several "calories" have been in use:
- the 15 °C calorie: $1\,\mathrm{cal}_{15} = 4.1855\,\mathrm{J}$ (value adopted by the CIPM in 1950);
- the IT (International Table) calorie: $1\,\mathrm{cal}_{\mathrm{IT}} = 4.1868\,\mathrm{J}$ (5th International Conference on the Properties of Steam, London, 1956);
- the thermochemical calorie: $1\,\mathrm{cal}_{\mathrm{th}} = 4.184\,\mathrm{J}$.

[j] The micron and its symbol, adopted by the CIPM in 1879 and repeated in Resolution 7 of the 9th CGPM (1948), were abolished by the 13th CGPM (1967–1968).

Table 1.2-16 Internationally adopted numerical values for selected quantities

Quantity	Symbol	Numerical value	Unit
Relative atomic mass [a] of ^{12}C	$A_r(^{12}\mathrm{C})$	12	
Molar mass constant	M_u	1×10^{-3}	kg/mol
Molar mass of ^{12}C	$M(^{12}\mathrm{C})$	12×10^{-3}	kg/mol
Conventional value of the Josephson constant [b]	$K_{\mathrm{J\text{-}90}}$	483 597.9	GHz/V
Conventional value of the von Klitzing constant [c]	$R_{\mathrm{K\text{-}90}}$	25 812.807	Ω
Standard atmosphere		101 325	Pa
Standard acceleration of free fall [d]	g_n	9.80665	$\mathrm{m/s^2}$

[a] The relative atomic mass $A_r(X)$ of a particle X with mass $m(X)$ is defined by $A_r(X) = m(X)/m_u$, where $m_u = m(^{12}\mathrm{C})/12 = M_u/N_A = 1\,\mathrm{u}$ is the atomic mass constant, M_u is the molar mass constant, N_A is the Avogadro number, and u is the (unified) atomic mass unit. Thus the mass of a particle X is $m(X) = A_r(X)\,\mathrm{u}$ and the molar mass of X is $M(X) = A_r(X)M_u$.

[b] This is the value adopted internationally for realizing representations of the volt using the Josephson effect.

[c] This is the value adopted internationally for realizing representations of the ohm using the quantum Hall effect.

[d] The value given was adopted by the 3rd General Conference on Weights and Measures (CGPM), 1903, and was the conventional value used to calculate the now obsolete unit kilogram force.

1.2.7 Some Energy Equivalents

In science and technology, energy is measured in many different units. Different units are used depending on the field of application, but owing to the different possible forms of the energy concerned, it is possible also to express the energy in terms of other quantities. All forms of the energy, however, are quantitatively related to one another and are therefore considered as being equivalent. Some of the most important equivalence relations are

$$E = eU = mc^2 = hc/\lambda = h\nu = kT \ .$$

These equations tell us that a given energy E, which is usually measured either in units of joule (J) or units of the Hartree energy ($E_H = 1$ hartree), can also be specified by giving a voltage U, a mass m, a wavelength λ, a frequency ν, or a temperature T. These equations contain, in addition to those variables, only well-known fundamental constants.

Table 1.2-17 gives the values of the energy equivalents of the joule and the hartree and for the SI units corresponding to the five quantities U, m, λ, ν, and T. The equivalents have been calculated on the basis of the 2002 CODATA adjustment of the values of the constants.

Table 1.2-17 Energy equivalents, expressed in the units joule (J), hartree (E_H), volt (V), kilogram (kg), (unified) atomic mass unit (u), reciprocal meter (m^{-1}), hertz (Hz), and kelvin (K)

	Unit			
Energy	**Joule**	**Hartree**	**Volt**	**Kilogram**
1 J	$(1\,J) = 1\,J$	$(1\,J) = 2.29371257(39) \times 10^{17}\,E_H$	$(1\,J) = 6.24150947(53) \times 10^{18}\,eV$	$(1\,J)/c^2 = 1.112650056 \times 10^{-17}\,kg$
1 E_H	$(1\,E_H) = 4.35974417(75) \times 10^{-18}\,J$	$(1\,E_H) = 1\,E_H$	$(1\,E_H) = 27.2113845(23)\,eV$	$(1\,E_H)/c^2 = 4.85086960(83) \times 10^{-35}\,kg$
1 eV	$(1\,eV) = 1.60217653(14) \times 10^{-19}\,J$	$(1\,eV) = 3.67493245(31) \times 10^{-2}\,E_H$	$(1\,eV) = 1\,eV$	$(1\,eV)/c^2 = 1.78266181(15) \times 10^{-36}\,kg$
1 kg	$(1\,kg)\,c^2 = 8.987551787 \times 10^{16}\,J$	$(1\,kg)\,c^2 = 2.06148605(35) \times 10^{34}\,E_H$	$(1\,kg)\,c^2 = 5.60958896(48) \times 10^{35}\,eV$	$(1\,kg) = 1\,kg$
1 u	$(1\,u)\,c^2 = 1.49241790(26) \times 10^{-10}\,J$	$(1\,u)\,c^2 = 3.423177686(23) \times 10^{7}\,E_H$	$(1\,u)\,c^2 = 931.494043(80) \times 10^{6}\,eV$	$(1\,u) = 1.66053886(28) \times 10^{-27}\,kg$
1 m^{-1}	$(1\,m^{-1})\,hc = 1.98644561(34) \times 10^{-25}\,J$	$(1\,m^{-1})\,hc = 4.556335252760(30) \times 10^{-8}\,E_H$	$(1\,m^{-1})\,hc = 1.23984191(11) \times 10^{-6}\,eV$	$(1\,m^{-1})\,h/c = 2.21021881(38) \times 10^{-42}\,kg$
1 Hz	$(1\,Hz)\,h = 6.6260693(11) \times 10^{-34}\,J$	$(1\,Hz)\,h = 1.519829846006(10) \times 10^{-16}\,E_H$	$(1\,Hz)\,h = 4.13566743(35) \times 10^{-15}\,eV$	$(1\,Hz)\,h/c^2 = 7.3724964(13) \times 10^{-51}\,kg$
1 K	$(1\,K)\,k = 1.3806505(24) \times 10^{-23}\,J$	$(1\,K)\,k = 3.1668153(55) \times 10^{-6}\,E_H$	$(1\,K)\,k = 8.617343(15) \times 10^{-5}\,eV$	$(1\,K)\,k/c^2 = 1.5361808(27) \times 10^{-40}\,kg$

	Unit			
Energy	**Atomic mass unit**	**Reciprocal meter**	**Hertz**	**Kelvin**
1 J	$(1\,J)/c^2 = 6.7005361(11) \times 10^{9}\,u$	$(1\,J)/hc = 5.03411720(86) \times 10^{24}\,m^{-1}$	$(1\,J)/h = 1.50919037(26) \times 10^{33}\,Hz$	$(1\,J)/k = 7.242963(13) \times 10^{22}\,K$
1 E_H	$(1\,E_H)/c^2 = 2.921262323(19) \times 10^{-8}\,u$	$(1\,E_H)/hc = 2.194746313705(15) \times 10^{7}\,m^{-1}$	$(1\,E_H)/h = 6.579683920721(44) \times 10^{15}\,Hz$	$(1\,E_H)/k = 3.1577465(55) \times 10^{5}\,K$
1 eV	$(1\,eV)/c^2 = 1.073544171(92) \times 10^{-9}\,u$	$(1\,eV)/hc = 8.06554445(69) \times 10^{5}\,m^{-1}$	$(1\,eV)/h = 2.41798940(21) \times 10^{14}\,Hz$	$(1\,eV)/k = 1.1604505(20) \times 10^{4}\,K$
1 kg	$(1\,kg) = 6.0221415(10) \times 10^{26}\,u$	$(1\,kg)\,c/h = 4.52443891(77) \times 10^{41}\,m^{-1}$	$(1\,kg)\,c^2/h = 1.35639266(23) \times 10^{50}\,Hz$	$(1\,kg)\,c^2/k = 6.509650(11) \times 10^{39}\,K$
1 u	$(1\,u) = 1\,u$	$(1\,u)\,c/h = 7.513006608(50) \times 10^{14}\,m^{-1}$	$(1\,u)\,c^2/h = 2.252342718(15) \times 10^{23}\,Hz$	$(1\,u)\,c^2/k = 1.0809527(19) \times 10^{13}\,K$
1 m^{-1}	$(1\,m^{-1})\,h/c = 1.3310250506(89) \times 10^{-15}\,u$	$(1\,m^{-1}) = 1\,m^{-1}$	$(1\,m^{-1})\,c = 299\,792\,458\,Hz$	$(1\,m^{-1})\,hc/k = 1.4387752(25) \times 10^{-2}\,K$
1 Hz	$(1\,Hz)\,h/c^2 = 4.439821667(30) \times 10^{-24}\,u$	$(1\,Hz)/c = 3.335640951 \times 10^{-9}\,m^{-1}$	$(1\,Hz) = 1\,Hz$	$(1\,Hz)\,h/k = 4.7992374(84) \times 10^{-11}\,K$
1 K	$(1\,K)\,k/c^2 = 9.251098(16) \times 10^{-14}\,u$	$(1\,K)\,k/hc = 69.50356(12)\,m^{-1}$	$(1\,K)\,k/h = 2.0836644(36) \times 10^{10}\,Hz$	$(1\,K) = 1\,K$

References

2.1 Bureau International des Poids et Mesures: *Le système international d'unités*, 7th edn. (Bureau International des Poids et Mesures, Sèvres 1998)

2.2 Organisation Intergouvernementale de la Convention du Mètre: *The International System of Units (SI), Addenda and Corrigenda to the 7th Edition* (Bureau International des Poids et Mesures, Sèvres 2000)

1.3. Rudiments of Crystallography

Crystallography deals basically with the question "Where are the atoms in solids?" The purpose of this section is to introduce briefly the basics of modern crystallography. The focus is on the description of periodic solids, which represent the major proportion of condensed matter. A coherent introduction to the formalism required to do this is given, and the basic concepts and technical terms are briefly explained. Paying attention to recent developments in materials research, we treat aperiodic, disordered, and amorphous materials as well. Consequently, besides the conventional three-dimensional (3-D) descriptions, the higher-dimensional crystallographic approach is outlined, and so is the atomic pair distribution function used to describe local phenomena. The section is concluded by touching on the basics of diffraction methods, the most powerful tool kit used by experimentalists dealing with structure at the atomic level in the solid state.

The structure of a solid material is very important, because the physical properties are closely related to the structure. In most cases solids are crystalline: they may consist of one single crystal, or be polycrystalline, consisting of many tiny single crystals in different orientations. All periodic crystals have a perfect translational symmetry. This leads to selection rules, which are very useful for the understanding of the physical properties of solids. Therefore, most textbooks on solid-state physics begin with some chapters on symmetry and structure. Today we know that other solids, which have no translational symmetry, also exist. These are amorphous materials, which have little order (in most cases restricted to the short-range arrangement of the atoms), and aperiodic crystals, which show perfect long-range order, but no periodicity – at least in 3-D space. In this chapter of the book, the basic concepts of crystallography – how the space of a solid can be filled with atoms – are briefly discussed. Readers who want to inform themselves in more detail about crystallography are referred to the classic textbooks [3.1–5].

Many crystalline materials, especially minerals and gems, were described more than 2000 years ago. The regular form of crystals and the existence of facets, which have fixed angles between them, gave rise to a belief that crystals were formed by a regular repetition of tiny, identical building blocks. After the discovery of X-rays by Röntgen, Laue investigated crystals in 1912 using these X-rays and detected interference effects caused by the periodic array of atoms. One year later, Bragg determined the crystal structures of alkali halides by X-ray diffraction.

Today we know that a crystal is a 3-D array of atoms or molecules, with various types of long-range order. A more modern definition is that all materials which show sharp diffraction peaks are crystalline. In this sense, aperiodic or quasicrystalline materials, as well as periodic materials, are crystals. A real crystal is never a perfect arrangement. Defects in the form of vacancies, dislocations, impurities, and other imperfections are often very important for the physical properties of a crystal. This aspect has been largely neglected in classical crystallography but is becoming more and more a topic of modern crystallographic investigations [3.6, 7].

As indicated in Table 1.3-1, condensed matter can be classified as either crystalline or amorphous. Both of these states and their formal subdivisions will be discussed in the following. The terms "matter", "structure", and "material" always refer to single-phase solids.

Table 1.3-1 Classification of solids

Condensed matter (solids)				
Crystalline materials				Amorphous materials
Periodic structures	Aperiodic structures			
	Modulated structures	Composite structures	Quasicrystals	

1.3.1 Crystalline Materials

1.3.1.1 Periodic Materials

Lattice Concept

A periodic crystal is described by two entities, the *lattice* and the *basis*. The (translational) lattice is a perfect geometrical array of points. All lattice points are equivalent and have identical surroundings. This lattice is defined by three fundamental translation vectors $\boldsymbol{a}, \boldsymbol{b}, \boldsymbol{c}$. Starting from an arbitrarily chosen origin of the lattice, any other lattice point can be reached by a translation vector $\boldsymbol{r}$ that satisfies

$$\boldsymbol{r} = u\boldsymbol{a} + v\boldsymbol{b} + w\boldsymbol{c} ,$$

where u, v, and w are arbitrary integers.

The lattice is an abstract mathematical construction; the description of the crystal is completed by attaching a set of atoms – the basis – to each lattice point. Therefore the crystal structure is formed by a lattice and a basis (see Fig. 1.3-1).

The parallelepiped that is defined by the axes $\boldsymbol{a}, \boldsymbol{b}, \boldsymbol{c}$ is called a *primitive cell* if this cell has the smallest volume out of all possible cells. It contains one lattice point per cell only (Fig. 1.3-2a). This cell is a type of unit cell which fills the space of the crystal completely under the application of the translation operations of the lattice, i. e. movements along the vectors $\boldsymbol{r}$.

Conventionally, the smallest cell with the highest symmetry is chosen. Crystal lattices can be transformed into themselves by translation along the fundamental vectors $\boldsymbol{a}, \boldsymbol{b}, \boldsymbol{c}$, but also by other symmetry operations. It can be shown that only onefold (rotation angle $\varphi = 2\pi/1$), twofold ($2\pi/2$), threefold ($2\pi/3$), fourfold ($2\pi/4$), and sixfold ($2\pi/6$) rotation axes are permissible. Other rotational axes cannot exist in a lattice, because they would violate the translational symmetry. For example, it is not possible to fill the space completely with a fivefold ($2\pi/5$) array of regular pentagons. Additionally, mirror planes and centers of inversion may exist. The restriction to high-symmetry cells may also lead to what is known as centering. Figure 1.3-2b illustrates a 2-D case. The centering types in 3-D are listed in Table 1.3-2.

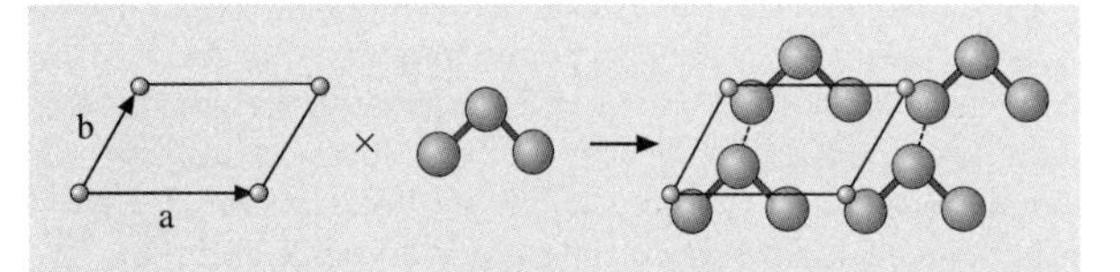

Fig. 1.3-1 A periodic crystal can be described as a convolution of a mathematical point lattice with a basis (set of atoms). *Open circles*, mathematical points; *filled circles*, atoms

Planes and Directions in Lattices

If one peers through a 3-D lattice from various angles, an infinity of equidistant planes can be seen. The position and orientation of such a crystal plane are determined by three points. It is easy to describe a plane if all three points lie on crystal axes (i. e. the directions of unit cell vectors); in this case only the intercepts need to be used. It is common to use *Miller indices* to de-

Table 1.3-2 Centering types for 3-D crystallographic unit cells

Symbol	Description	Points per unit cell
P	No centering (primitive)	1
I	Body-centered (*innenzentriert*)	2
F	All-face-centered	4
S; *A*, *B*, *C* in specific cases	One-face-centered (*seitenzentriert*); ($\boldsymbol{b}, \boldsymbol{c}$), ($\boldsymbol{a}, \boldsymbol{c}$), and ($\boldsymbol{a}, \boldsymbol{b}$), respectively, in specific cases	2
R	Hexagonal cell, rhombohedrally centered	3

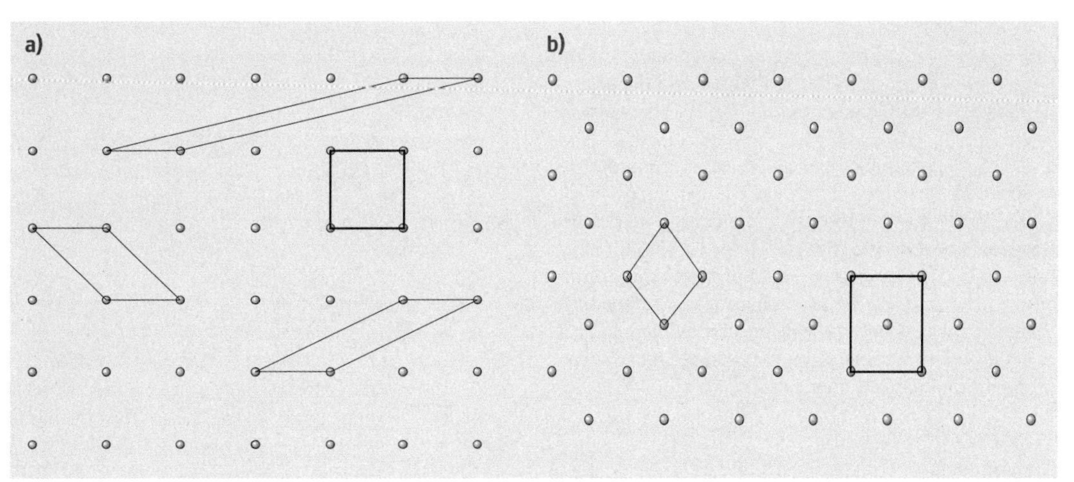

Fig. 1.3-2a,b Possible primitive and centered cells in 2-D lattices. *Open circles* denote mathematical points. (**a**) In this lattice, the conventional cell is the *bold square cell* because of its highest symmetry, 4*mm*. (**b**) Here, convention prefers 90° angles: a centered cell of symmetry 2*mm* is chosen. It contains two lattice points and is twice the area of the primitive cell

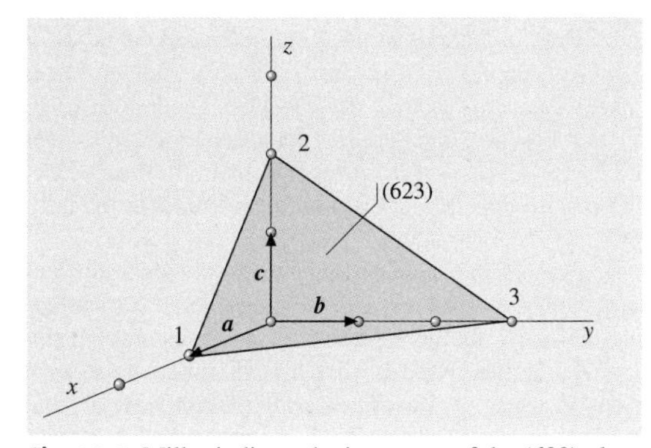

Fig. 1.3-3 Miller indices: the intercepts of the (623) plane with the coordinate axes

scribe lattice planes. These indices are determined as follows:

1. For the plane of interest, determine the intercepts x, y, z of the crystal axes $\boldsymbol{a}, \boldsymbol{b}, \boldsymbol{c}$.
2. Express the intercepts in terms of the basic vectors $\boldsymbol{a}, \boldsymbol{b}, \boldsymbol{c}$ of the unit cell, i. e. as x/a, y/b, z/c (where $a = |\boldsymbol{a}|, \ldots$).
3. Form the reciprocals a/x, b/y, c/z.
4. Reduce this set to the smallest integers h, k, l. The result is written (hkl).

The distance from the origin to the plane (hkl) inside the unit cell is the interplanar spacing d_{hkl}. Negative intercepts, leading to negative Miller indices, are written as $\bar{h}$. Figure 1.3-3 shows a (623) plane and its construction.

A direction in a crystal is given as a set of three integers in square brackets $[uvw]$; u, v, and w correspond to the above definition of the translation vector $\boldsymbol{r}$. A direction in a cubic crystal can be described also by Miller indices, as a plane can be defined by its normal. The indices of a direction are expressed as the smallest integers which have the same ratio as the components of a vector (expressed in terms of the axis vectors $\boldsymbol{a}, \boldsymbol{b}, \boldsymbol{c}$) in that direction. Thus the sets of integers 1, 1, 1 and 3, 3, 3 represent the same direction in a crystal, but the indices of the direction are [111] and not [333]. To give another example, the x axis of an orthogonal x, y, z coordinate system has Miller indices [100]; the plane perpendicular to this direction has indices (100).

For all crystals, except for the hexagonal system, the Miller indices are given in a three-digit system in the form (hkl). However, for the hexagonal system, it is common to use four digits $(hkil)$. The four-digit hexagonal indices are based on a coordinate system containing four axes. Three axes lie in the basal plane of the hexagon, crossing at angles of 120°: $\boldsymbol{a}$, $\boldsymbol{b}$, and $-(\boldsymbol{a}+\boldsymbol{b})$. As the third vector in the basal plane can be expressed in terms of $\boldsymbol{a}$ and $\boldsymbol{b}$, the index can be expressed in terms of h and k: $i = -(h+k)$. The fourth axis is the $\boldsymbol{c}$ axis normal to the basal plane.

Crystal Morphology

The regular facets of a crystal are planes of the type described above. Here, the lattice architecture of the crystal is visible macroscopically at the surface. Figure 1.3-4 shows some surfaces of a cubic crystal. If the crystal had the shape or morphology of a cube, this would be described by the set of facets $\{(100), (010), (001), (\bar{1}00), (0\bar{1}0), (00\bar{1})\}$. An octahedron would be described by $\{(111), (\bar{1}11), (1\bar{1}1), (11\bar{1}), (\bar{1}\bar{1}\bar{1}), (1\bar{1}\bar{1}), (\bar{1}1\bar{1}), (\bar{1}\bar{1}1)\}$. The morphology of a crystalline material may be of technological interest (in relation to the bulk density, flow properties, etc.) and can be influenced in various ways, for example by additives during the crystallization process.

The 32 Crystallographic Point Groups

The symmetry of the space surrounding a lattice point can be described by the point group, which is a set of symmetry elements acting on the lattice. The crystallographic symbols for the symmetry elements of point groups compatible with a translational lattice are the rotation axes $1, 2, 3, 4$, and 6, mirror planes m, and the center of inversion $\bar{1}$. Figure 1.3-5 illustrates, as an example, the point group $2/m$. The "2" denotes a twofold axis perpendicular ("/") to a mirror plane "m". Note that this combination of 2 and m implies, or generates automatically, an inversion center $\bar{1}$. We have used the Hermann–Mauguin notation here; however, point groups of isolated molecules are more often denoted by the Schoenflies symbols. For a translation list, see Table 1.3-3.

No crystal can have a higher point group symmetry than the point group of its lattice, called the *holohedry*. In accordance with the various rotational symmetries,

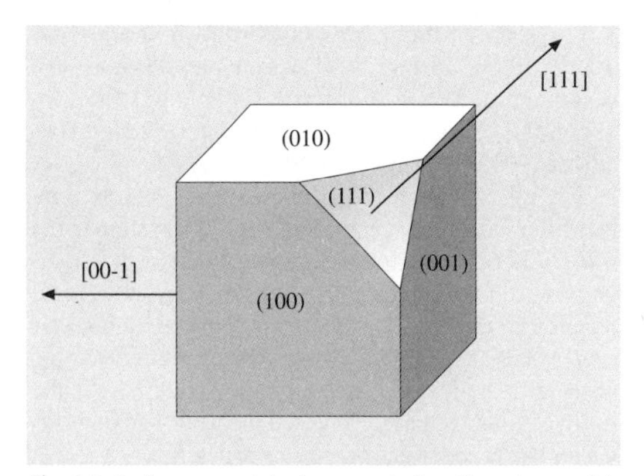

Fig. 1.3-4 Some crystal planes and directions in a cubic crystal, and their Miller indices

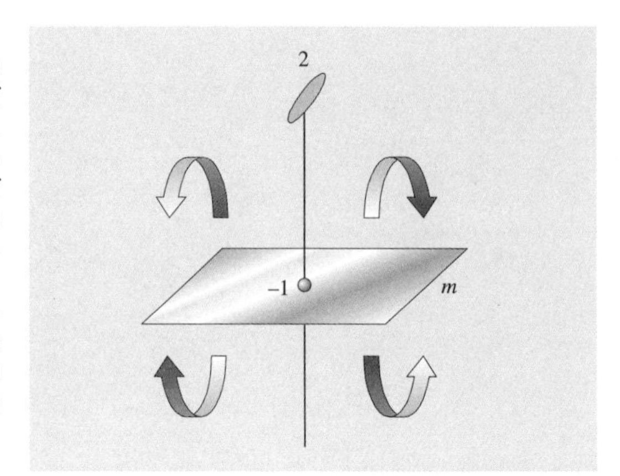

Fig. 1.3-5 The point group $2/m$ (C_{2h}). Any object in space can be rotated by $\varphi = 2\pi/2$ around the twofold rotational axis 2 and reflected by the perpendicular mirror plane m, generating identical copies. The inversion center $\bar{1}$ is implied by the coupling of 2 and m

there are seven crystal systems (see Table 1.3-3), and the seven holohedries are $\bar{1}$, $2/m$, mmm, $4/mmm$, $\bar{3}m$, $6/mmm$, and $m\bar{3}m$. Other, less symmetric, point groups are also compatible with these lattices, leading to a total number of 32 crystallographic point groups (see Table 1.3-4). A lower symmetry than the holohedry can be introduced by a less symmetric basis in the unit cell.

Since $\bar{3}m$ and $6/mmm$ are included in the same point lattice, they are sometimes subsumed into the hexagonal crystal family. So there are seven crystal systems but six crystal families. Note further that rhombohedral symmetry is a special case of centering (R-centering) of the trigonal crystal system and offers two equivalent possibilities for selecting the cell parameters: hexagonal or rhombohedral axes (see Table 1.3-4 again).

It can be shown that in 3-D there are 14 different periodic ways of arranging identical points. These 14 3-D periodic point lattices are called the (translational) Bravais lattices and are shown in Fig. 1.3-6. Table 1.3-4 presents data related to some of the crystallographic terms used here. The 1-D and 2-D space groups can be classified analogously but are omitted here.

The 230 Crystallographic Space Groups

Owing to the 3-D translational periodicity, symmetry operations other than point group operations are possible in addition: these are *glide planes* and *screw axes*. A glide plane couples a mirror operation and a translational shift. The symbols for glide planes

Table 1.3-3 The 32 crystallographic point groups: translation list from the Hermann–Mauguin to the Schoenflies notation

Crystal system	Hermann–Mauguin symbol	Schoenflies symbol	Crystal system	Hermann–Mauguin symbol	Schoenflies symbol
Triclinic	1	C_1	Trigonal	3	C_3
	$\bar{1}$	C_i		$\bar{3}$	C_{3i}
Monoclinic	2	C_2		32	D_3
	m	C_s		$3m$	C_{3v}
	$2/m$	C_{2h}		$\bar{3}m$	D_{3d}
Orthorhombic	222	D_2	Hexagonal	6	C_6
	$mm2$	C_{2v}		$\bar{6}$	C_{3h}
	mmm	D_{2h}		$6/m$	C_{6h}
Tetragonal	4	C_4		622	D_6
	$\bar{4}$	S_4		$6mm$	C_{6v}
	$4/m$	C_{4h}		$\bar{6}2m$	D_{3h}
	422	D_4		$6/mmm$	D_{6h}
	$4mm$	C_{4v}	Cubic	23	T
	$\bar{4}2m$	D_{2d}		$m\bar{3}$	T_h
	$4/mmm$	D_{4h}		432	O
				$\bar{4}3m$	T_d
				$m\bar{3}m$	O_h

are a, b, and c for translations along the lattice vectors $\boldsymbol{a}, \boldsymbol{b}$, and $\boldsymbol{c}$, respectively, and n and d for some special lattice vector combinations. A screw axis is always parallel to a rotational axis. The symbols are $2_1, 3_1, 3_2, 4_1, 4_2, 4_3, 6_1, 6_2, 6_3, 6_4$, and 6_5, where, for example, 6_3 means a rotation through an angle $\varphi = 2\pi/6$ followed by a translation of $3/6\,(= 1/2)$ of a full translational period along the sixfold axis.

Thus the combination of 3-D translational and point symmetry operations leads to an infinite number of sets of symmetry operations. Mathematically, each of these sets forms a group, and they are called *space groups*. It can be shown that all possible periodic crystals can be described by only 230 space groups. These 230 space groups are described in tables, for example the *International Tables for Crystallography* [3.8].

In this formalism, a conventional space group symbol reflects the symmetry elements, arranged in the order of standardized *blickrichtungen* (symmetry directions). We shall confine ourselves here to explain one instructive example: $P4_2/mcm$, space group number 132 [3.8]. The full space group symbol is $P\,4_2/m\,2/c\,2/m$. The mean-

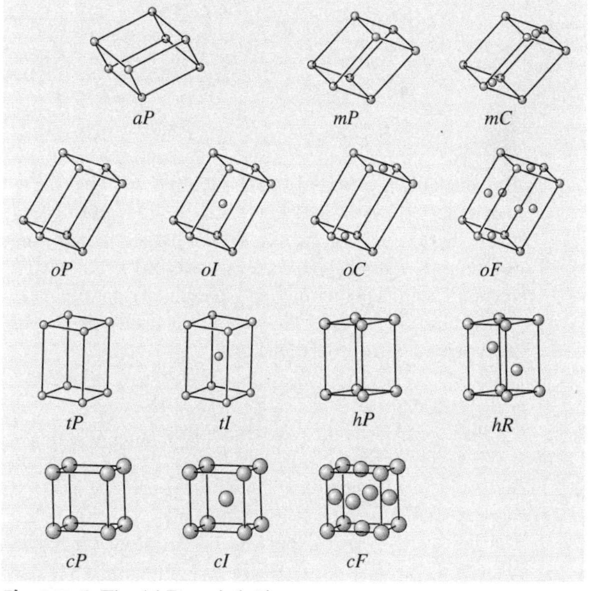

Fig. 1.3-6 The 14 Bravais lattices

Table 1.3-4 Crystal families, crystal systems, crystallographic point groups, conventional coordinate systems, and Bravais lattices in three dimensions. Lattice point symmetries (holohedries) are given in **bold**

Crystal family	Symbol	Crystal system	Crystallographic point groups	No. of space groups	Conventional coordinate system		Bravais lattice (Pearson symbol)
					Restrictions on cell parameters	Parameters to be determined	
Triclinic (*a*northic)	*a*	Triclinic	1, $\mathbf{\bar{1}}$	2	None	$a, b, c,$ α, β, γ	aP
*m*onoclinic	*m*	Monoclinic	2, m, $\mathbf{2/m}$	13	Setting with b unique: $\alpha = \gamma = 90°$	$a, b, c,$ β	mP, mS (mC, mA, mI)
					Setting with c unique: $\alpha = \beta = 90°$	$a, b, c,$ γ	mP, mS (mA, mB, mI)
*o*rthorhombic	*o*	Orthorhombic	222, $mm2$, $\mathbf{mmm}$	59	$\alpha = \beta = \gamma = 90°$	a, b, c	oP,oS, (oC, oA, oB) oI, oF
*t*etragonal	*t*	Tetragonal	$4, \bar{4}, 4/m,$ $422, 4mm,$ $\bar{4}2m$, $\mathbf{4/mmm}$	68	$a = b$ $\alpha = \beta = \gamma = 90°$	a, c	tP, tI
*h*exagonal	*h*	Trigonal	$3, \bar{3},$ $32, 3m$, $\mathbf{\bar{3}m}$	18	$a = b,$ $\alpha = \beta = 90°$ $\gamma = 120°$ (hexagonal axes)	a, c	hP
				7	$a = b = c$ $\alpha = \beta = \gamma \neq 90°$ (rhombohedral axes)	a, α	hR
		Hexagonal	$6, \bar{6}, 6/m,$ $622, 6mm,$ $\bar{6}2m$, $\mathbf{6/mmm}$	27	$a = b,$ $\alpha = \beta = 90°$ $\gamma = 120°$	a, c	hP
cubic	*c*	Cubic	$23, m\bar{3},$ $432, \bar{4}3m,$ $\mathbf{m\bar{3}m}$	36	$a = b = c$ $\alpha = \beta = \gamma = 90°$	a	cP, cI,cF

ing of the symbols is the following: P denotes a primitive Bravais lattice. It belongs to the tetragonal crystal system indicated by 4. Along the first standard *blickrichtung* [001] there is a 4_2 screw axis with a perpendicular mirror plane m. Along [100] there is a twofold rotation axis, named 2, with a perpendicular glide plane c parallel to $\boldsymbol{c}$. Third, along [110] there is a twofold rotation axis 2, with a perpendicular mirror plane m.

Decoration of the Lattice with the Basis

At this point we have to recall that in a real crystal structure we have not only the lattice, but also the basis. In [3.8], there are standardized sets of general and special positions (i. e. coordinates x, y, z) within the unit cell (Wyckoff positions). An atom placed in a general position is transformed into more than one atom by the action of all symmetry operators of the respective space group. Special positions are located on special points which are mapped onto themselves by one or more symmetry operations – for example a position in a mirror plane or exactly on a rotational axis. Reference [3.8] also provides information about symmetry relations between individual space groups (group–subgroup relations). These are often useful for describing relationships between crystal structures and for describing phase transitions of materials.

The use of the space group allows us to further reduce the basis to the *asymmetric unit*: this is the minimal set of atoms that needs to be given so that the whole crystal structure can be generated via the symmetry of the space group. This represents the main power of a crystallographically correct description of a material: just some 10 parameters are sufficient to describe an ensemble of some 10^{23} atoms.

Thus, a crystallographically periodic structure of a material is unambiguously characterized by

- the cell parameters;
- the space group;
- the coordinates of the atoms (and their chemical type) in the asymmetric unit;
- the occupation and thermal displacement factors of the atoms in the asymmetric unit.

For an example, the reader is referred to the crystallographic description of the spinel structure of $MgAl_2O_4$ given below under the heading "Structure Types".

To complete the information on space group symmetries given here, periodic magnetic materials should also be mentioned. Magnetic materials contain magnetic moments carried by atoms in certain positions in the unit cell. If we take into account the magnetic moments in the description of the structure, the classification by space groups (the 230 "gray" groups, described above) has to be extended to 1651 the "black and white", or Shubnikov, groups [3.9]. A magnetic periodic structure is then characterized by

- the crystallographic structure;
- the Shubnikov group;
- the cell parameters of the magnetic unit cell;
- the coordinates of the atoms carrying magnetic moments (the asymmetric unit in the magnetic unit cell);
- the magnitude and direction of the magnetic moments on these atoms.

Structure Types

It is useful to classify the crystal structures of materials by the assignment of *structure types*. The structure type is based on a representative crystal structure, the parameters of which describe the essential crystallographic features of other materials of the same type. As an example, we consider the structure of the spinel oxides AB_2O_4. The generic structure type is $MgAl_2O_4$, cF56. The *Pearson symbol*, here cF56, denotes the cubic crystal family and a face-centered Bravais lattice with 56 atoms per unit cell (see Table 1.3-5 and also the last column in Table 1.3-4).

Regarding the free parameters, for example a, the notation 8.174(1) in Table 1.3-5 means 8.174 ± 0.001. The chemical formula and the unit cell contents can easily be calculated from the site multiplicities (given by the Wyckoff positions) and the occupancies. So can the (crystallographic) density, using the appropriate atomic masses.

There is a huge variety of other materials belonging to the same structure type as in this example. The only parameters that differ (slightly) are the numerical value of a, the types of atoms in the positions, the numerical value of the parameter x for Wyckoff position 32e, and the occupancies. Thus, for example, the crystal structure of the iron sulfide Fe_3S_4 can be characterized in its essential features via the information that it belongs to the same structure type.

1.3.1.2 Aperiodic Materials

In addition to the crystalline periodic state of matter, a class of materials exists that lacks 3-D translational symmetry and is called *aperiodic*. Aperiodic materials cannot be described by any of the 230 space groups mentioned above. Nevertheless, they show another type of long-range order and are therefore included in the term "crystal". This notion of long-range order is the major feature that distinguishes crystals from amorphous materials. Three types of aperiodic order may

Table 1.3-5 Complete crystallographic parameter set for $MgAl_2O_4$, spinel structure type

Material	$MgAl_2O_4$				
Structure type	$MgAl_2O_4$, spinel				
Pearson symbol	cF56				
Space group	$Fd\bar{3}m$ (No. 227)				
a (Å)	8.174(1)				
Atom	**Wyckoff position**	**x**	**y**	**z**	**Occupancy**
Mg	8a	0	0	0	1.0
Al	16d	5/8	5/8	5/8	1.0
O	32e	0.3863(2)	x	x	1.0

be distinguished, namely modulated structures, composite structures, and quasicrystals. All aperiodic solids exhibit an essentially discrete diffraction pattern and can be described as atomic structures obtained from a 3-D section of a n-dimensional (n-D) ($n > 3$) periodic structure.

Modulated Structures

In a modulated structure, periodic deviations of the atomic parameters from a reference or basic structure are present. The basic structure can be understood as a periodic structure as described above. Periodic deviations of one or several of the following atomic parameters are superimposed on this basic structure:

- atomic coordinates;
- occupancy factors;
- thermal displacement factors;
- orientations of magnetic moments.

Let the period of the basic structure be a and the modulation wavelength be λ; the ratio a/λ may be (1) a rational or (2) an irrational number (Fig. 1.3-7). In case (1), the structure is commensurately modulated; we observe a qa superstructure, where $q = 1/\lambda$. This superstructure is periodic. In case (2), the structure is incommensurately modulated. Of course, the experimental distinction between the two cases is limited by the finite experimental resolution. q may be a function of external variables such as temperature, pressure, or chemical composition, i. e. $q = f(T, p, X)$, and may adopt a rational value to result in a commensurate "lock-in" structure. On the other hand, an incommensurate charge-density wave may exist; this can be moved through a basic crystal without changing the internal energy U of the crystal.

When a 1-D basic structure and its modulation function are combined in a 2-D hyperspace $\boldsymbol{R} = \boldsymbol{R}^{\text{parallel}} \oplus \boldsymbol{R}^{\text{perpendicular}}$, periodicity on a 2-D lattice results. The real atoms are generated by the intersection of the 1-D physical (external, parallel) space $\boldsymbol{R}^{\text{parallel}}$ with the hyperatoms in the complementary 1-D internal space $\boldsymbol{R}^{\text{perpendicular}}$. In the case of a modulated structure, the hyperatoms have the shape of the sinusoidal modulation function in $\boldsymbol{R}^{\text{perpendicular}}$.

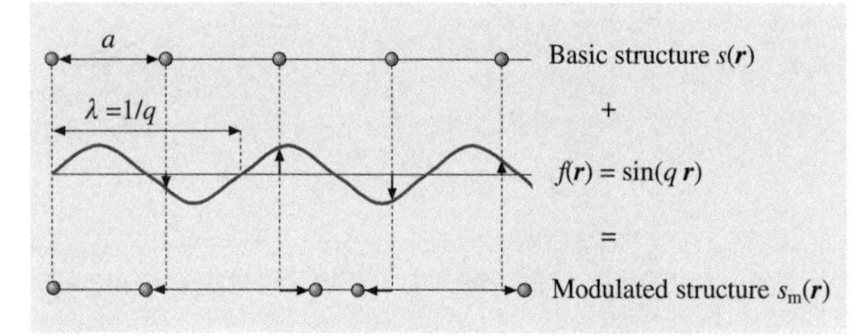

Fig. 1.3-7 A 1-D modulated structure $s_m(\boldsymbol{r})$ can be described as a sum of a basic structure $s(\boldsymbol{r})$ and a modulation function $f(\boldsymbol{r})$ of its atomic coordinates. If a/λ is irrational, the structure is incommensurately modulated. *Circles* denote atoms

Figure 1.3-8 illustrates this construction. We have to choose a basis $(\boldsymbol{a}_1, \boldsymbol{a}_2)$ in $\boldsymbol{R}$ where the slope of $\boldsymbol{a}_1$ with respect to $\boldsymbol{R}^{\text{parallel}}$ corresponds to the length of the modulation λ.

It is clear that real atomic structures are always manifestations of matter in 3-D real, physical space. The cutting of the 2-D hyperspace to obtain real 1-D atoms illustrated in Fig. 1.3-8 may serve as an instructive basic example of the concept of higher-dimensional (n-D, $n > 3$) crystallography. The concept is also called a superspace description; it applies to all aperiodic structures and provides a convenient finite set of variables that can be used to compute the positions of all atoms in the real 3-D structure.

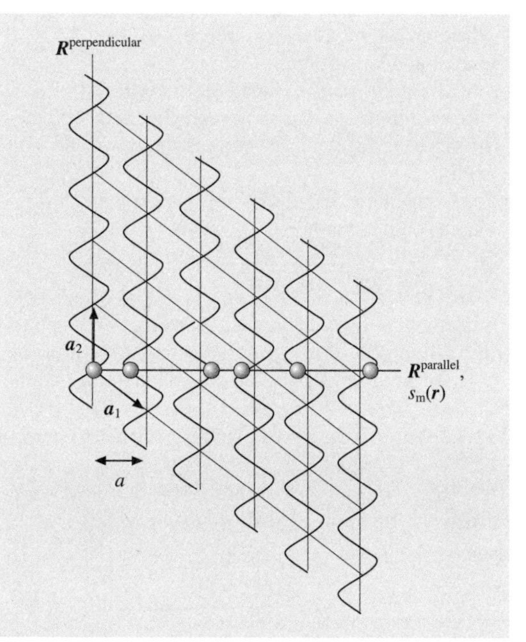

Fig. 1.3-8 2-D hyperspace description of the example of Fig. 1.3-7. The basis of the hyperspace $\boldsymbol{R} = \boldsymbol{R}^{\text{parallel}} \oplus \boldsymbol{R}^{\text{perpendicular}}$ is $(\boldsymbol{a}_1, \boldsymbol{a}_2)$; the slope of $\boldsymbol{a}_1$ with respect to $\boldsymbol{R}^{\text{parallel}}$ is proportional to λ. Atoms of the modulated structure $s_m(\boldsymbol{r})$ occur in the physical space $\boldsymbol{R}^{\text{parallel}}$ and are represented by *circles*

The modulation may occur in one, two, or three directions of the basic structure, yielding 1-D, 2-D, or 3-D modulated structures. If we introduce one additional dimension per modulation vector (the direction $\boldsymbol{r}$ that the modulation corresponding to λ runs along), these structures can be described as periodic in 4-D, 5-D, or 6-D superspace, respectively.

Composite Structures

Composite crystals are crystalline structures that consist of two or more periodic substructures, each one having its own 3-D periodicity to a first approximation. The symmetry of each of these subsystems is characterized by one of the 230 space groups. However, owing to their mutual interaction, the true structure consists of a collection of incommensurately modulated subsystems. All known composite structures to date have at least one lattice direction in common and consist of a maximum of three substructures. There are three main classes:

- channel structures;
- columnar packings;
- layer packings.

These composite structures are also known as intergrowth or host–guest structures. Figure 1.3-9 illustrates an example of a host with channels along $\boldsymbol{a}$, in which atoms of the substructure with a periodicity $\lambda\boldsymbol{a}$ reside as a guest.

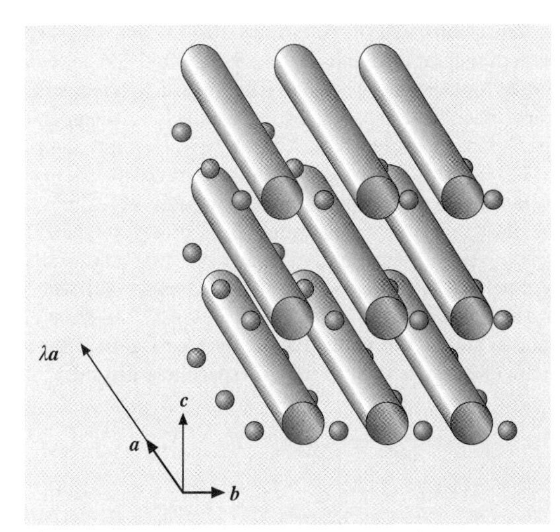

Fig. 1.3-9 Host–guest channel structure. The guest atoms reside in channels parallel to $\boldsymbol{a}$, with a periodicity $\lambda\boldsymbol{a}$

The higher-dimensional n-D formalism ($n > 3$) used to describe composite structures is essentially the same as that which applies to modulated structures.

Quasicrystals

Quasicrystals represent the third type of aperiodic materials. Quasiperiodicity may occur in one, two, or three dimensions of physical space and is associated with special irrational numbers such as the golden mean $\tau = (1+\sqrt{5})/2$, and $\xi = 2+\sqrt{3}$. The most remarkable feature of quasicrystals is the appearance of noncrystallographic point group symmetries in their diffraction patterns, such as $8/mmm$, $10/mmm$, $12/mmm$, and $2/m\bar{3}\bar{5}$. The golden mean is related to fivefold symmetry via the relation $\tau = 2\cos(\pi/5)$; τ can be considered as the "most irrational" number, since it is the irrational number that has the worst approximation by a truncated continued fraction,

$$\tau = 1 + \cfrac{1}{1+\cfrac{1}{1+\cfrac{1}{1+\cfrac{1}{1+\cfrac{1}{1+\ldots}}}}} .$$

This might be a reason for the stability of quasiperiodic systems where τ plays a role. A prominent 1-D example is the Fibonacci sequence, an aperiodic chain of short and long segments S and L with lengths S and L, where the relations $L/S = \tau$ and $L+S = \tau L$ hold. A Fibonacci chain can be constructed by the simple substitution or inflation rule L $\rightarrow$ LS and S $\rightarrow$ L (Table 1.3-6, Fig. 1.3-10). Materials quasiperiodically modulated in 1-D along one direction may occur. Again, their structures are readily described using the superspace formalism as above.

The Fibonacci sequence can be used to explain the idea of a periodic rational approximant. If the sequence ...LSLLSLSLS... represents a quasicrystal, then the

Table 1.3-6 Generation of the Fibonacci sequence using the inflation rule L $\rightarrow$ LS and S $\rightarrow$ L. The ratio F_{n+1}/F_n tends towards τ for $n \rightarrow \infty$. F_n is a Fibonacci number; $F_{n+1} = F_n + F_{n-1}$. The sequence starts with $F_0 = 0$, $F_1 = 1$

Sequence	n	F_{n+1}/F_n
L	1	$1/1 = 1$
LS	2	$2/1 = 2$
LSL	3	$3/2 = 1.5$
LSLLS	4	$5/3 = 1.66666\ldots$
LSLLSLSL	5	$8/5 = 1.6$
...		
... LSLLSLSLS ...	∞	$\tau = 1.61803\ldots$

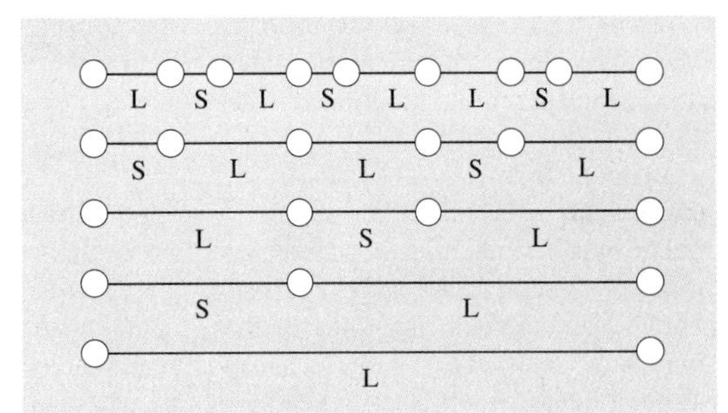

Fig. 1.3-10 1-D Fibonacci sequence. Moving downwards corresponds to an inflation of the self-similar chains, and moving upwards corresponds to a deflation

periodic sequence ...LSLSLSLSLS..., consisting only of the word LS, is its 2/1 approximant (Table 1.3-6). In real systems, such approximants often exist as large-unit-cell (periodic!) structures with atomic arrangements locally very similar to those in the corresponding quasicrystal. When described in terms of superspace, they would result via cutting with a rational slope, in the above example $2/1 = 2$, instead of $\tau = 1.6180\ldots$.

To date, all known 2-D quasiperiodic materials exhibit noncrystallographic diffraction symmetries of $8/mmm$, $10/mmm$, or $12/mmm$. The structures of these materials are called octagonal, decagonal, and dodecagonal structures, respectively. Quasiperiodicity is present only in planes stacked along a perpendicular periodic direction. To index the lattice points in a plane, four basis vectors $\boldsymbol{a}_1, \boldsymbol{a}_2, \boldsymbol{a}_3, \boldsymbol{a}_4$ are needed; a fifth one, $\boldsymbol{a}_5$, describes the periodic direction. Thus a 5-D hypercrystal is appropriate for describing the solid periodically. In an analogous way to the 230 3-D space groups, the 5-D superspace groups (e.g. $P10_5/mmc$) provide

- the multiplicity and Wyckoff positions;
- the site symmetry;
- the coordinates of the hyperatoms.

Again, the quasiperiodic structure in 3-D can be obtained from an intersection with the external space.

On the atomic scale, these quasicrystals consist of units of some 100 atoms, called clusters. These clusters, of point symmetry $8/mmm$, $10/mmm$, or $12/mmm$

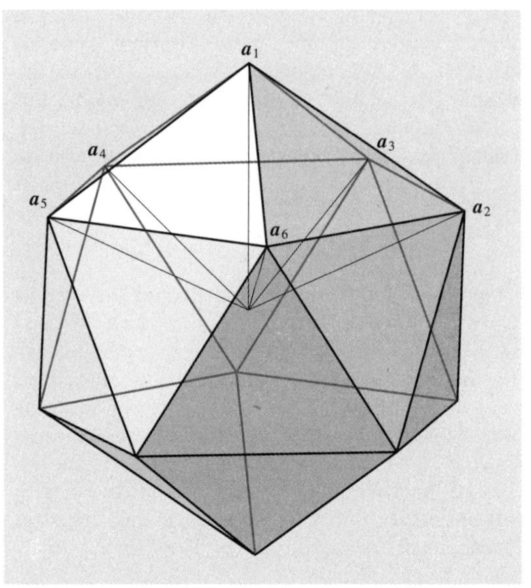

Fig. 1.3-11 Unit vectors $\boldsymbol{a}_1, \ldots, \boldsymbol{a}_6$ of an icosahedral lattice

(or less), are fused, may interpenetrate partially, and can be considered to decorate quasiperiodic tilings. In a diffraction experiment, their superposition leads to an overall noncrystallographic symmetry. There are a number of different tilings that show such noncrystallographic symmetries. Figure 1.3-12 depicts four of them, as examples of the octagonal, decagonal, and dodecagonal cases.

Icosahedral quasicrystals are also known. In 3-D, the icosahedral diffraction symmetry $2/m\bar{3}\bar{5}$ can be observed for these quasicrystals. Their diffraction patterns can be indexed using six integers, leading to a 6-D superspace description (see Fig. 1.3-11). On the atomic scale in 3-D, in physical space, clusters of some 100 atoms are arranged on the nodes of 3-D icosahedral tilings; the clusters have an icosahedral point group symmetry or less, partially interpenetrate, and generate an overall symmetry $2/m\bar{3}\bar{5}$. Many of their structures are still waiting to be determined completely. Figure 1.3-13 shows the two golden rhombohedra and the four Danzer tetrahedra that can be used to tile 3-D space icosahedrally.

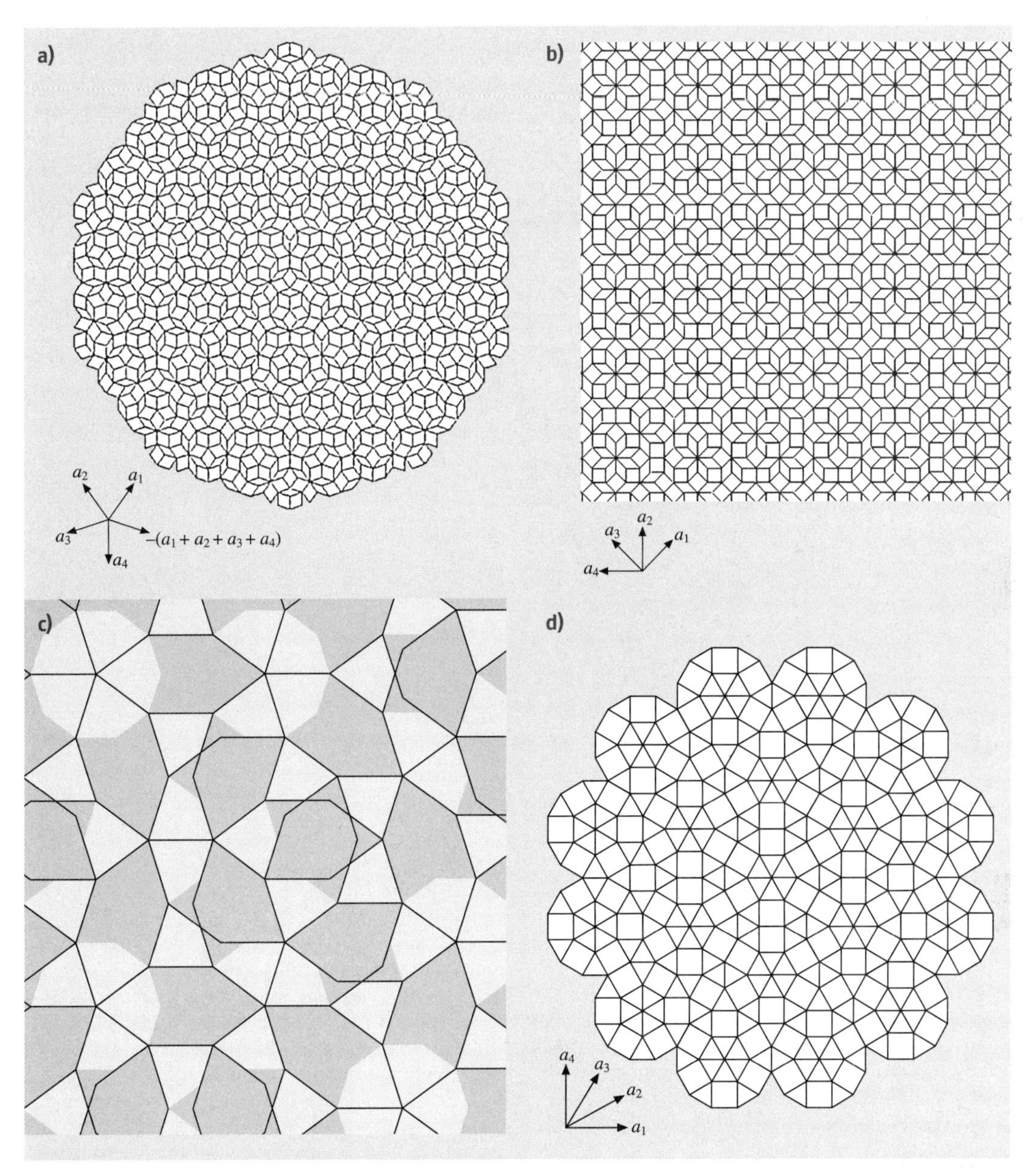

Fig. 1.3-12a–d Some 2-D quasiperiodic tilings; the corresponding four basis vectors $\boldsymbol{a}_1, \ldots, \boldsymbol{a}_4$ are shown. Linear combinations of $\boldsymbol{r} = \sum_i u_i \boldsymbol{a}_i$ reach all lattice points. (**a**) Penrose tiling with local symmetry $5mm$ and diffraction symmetry $10mm$, (**b**) octagonal tiling with diffraction symmetry $8mm$, (**c**) Gummelt tiling with diffraction symmetry $10mm$, and (**d**) dodecagonal Stampfli-type tiling with diffraction symmetry $12mm$

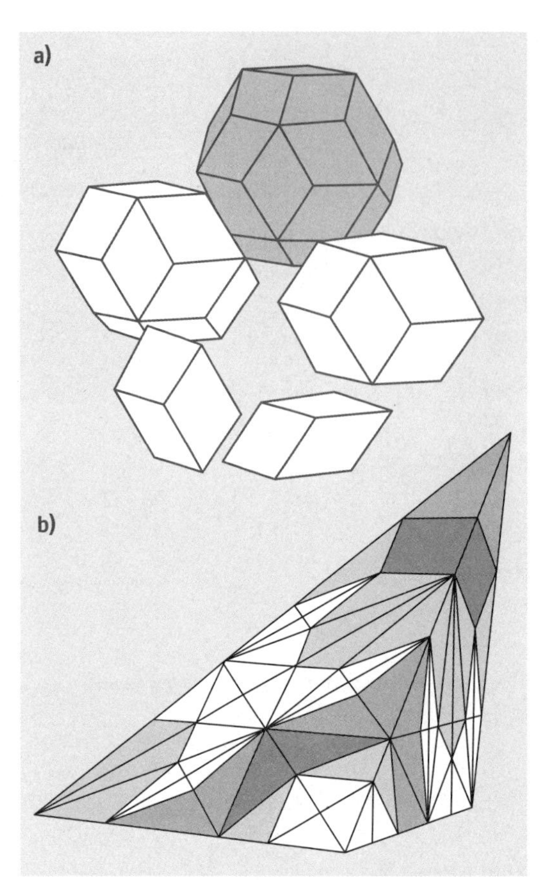

Fig. 1.3-13 Icosahedral tilings. (**a**) The two golden rhombohedra (*bottom*) can be used to form icosahedral objects (the rhombic triacontahedron with point symmetry $m\bar{3}\bar{5}$ shown in *gray*). (**b**) Danzer's {ABCK} tiling: three inflation steps for prototile A

1.3.2 Disorder

In between the ideal crystalline and the purely amorphous states, most real crystals contain degrees of disorder. Two types of statistical disorder have to be distinguished: chemical disorder and displacive disorder (Fig. 1.3-14). Statistical disorder contributes to the entropy S of the solid and is manifested by diffuse scattering in diffraction experiments. It may occur in both periodic and aperiodic materials.

Chemical Disorder

Chemical disorder is observed, for example, in the case of solid solutions, say of B in A, or $A_{1-x}B_x$ for short. Here, an average crystal structure exists. On the crystallographic atomic positions, different atomic species (the chemical elements A and B) are distributed randomly. Generally, the cell parameter a varies with x. For $x = 0$ or 1, the pure end member is present. A linear variation of $a(x)$ is predicted by Vegard's law. On the atomic scale, however, differences in the local structure, are present owing to the different contacts A–A, B–B, and A–B. These differences are usually represented by enlarged displacement factors, but can be investigated by analyzing the pair distribution function $G(r)$. $G(r)$ represents the probability of finding any atom at a distance r from any other atom relative to an average density. Chemical disorder can also occur on only one or a few of the crystallographically different atomic positions (e.g. $A(X_{1-x}Y_x)_2$). This type of disorder is often intrinsic to a material and may be temperature-dependent.

Displacive Disorder

The displacive type of disorder can be introduced by the presence of voids or vacancies in the structure or may exist for other reasons. Vacancies can be an important feature of a material: for example, they may leading to ionic conductivity or influence the mechanical properties.

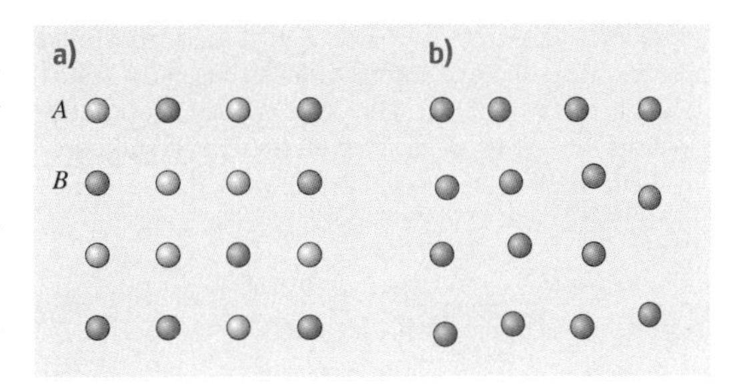

Fig. 1.3-14 Schematic sketch of (a) chemical and (b) displacive disorder

1.3.3 Amorphous Materials

The second large group of condensed matter is classified as the amorphous or glassy state. No long-range order is observed. The atoms are more or less statistically distributed in space, but a certain short-range order is present.

This short-range·order is reflected in the certain average coordination numbers or average coordination geometries. If there are strong (covalent) interactions between neighboring atoms, similar basic units may occur, which are in turn oriented randomly with respect to each other. The SiO_4 tetrahedron in silicate glasses is a well-known example. In an X-ray diffraction experiment on an amorphous solid, only isotropic diffuse scattering is observed. From this information, the radial atomic pair distribution function (Fig. 1.3-15) can be obtained. This function $G(r)$ can be interpreted as the probability of finding any atom at a distance r from any other atom relative to an average density.

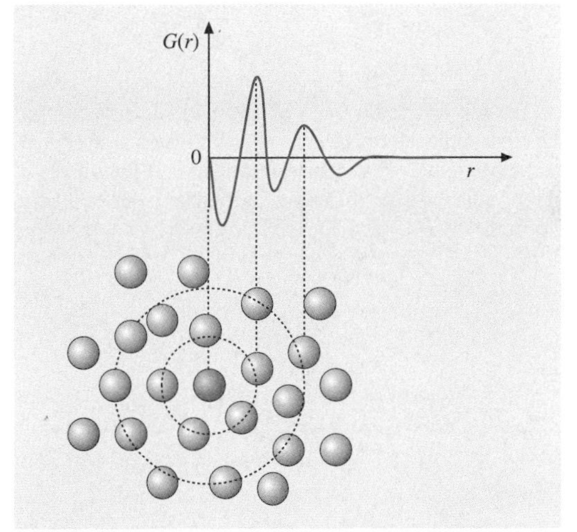

Fig. 1.3-15 Radial atomic pair distribution function $G(r)$ of an amorphous material. Its shape can be deduced from diffuse scattering

1.3.4 Methods for Investigating Crystallographic Structure

So far, we have been dealing with the formal description of solids. To conclude this chapter, the tool kit that an experimentalist needs to obtain structural information about a material in front of him/her will be briefly described.

The major technique used to derive the atomic structure of solids is the diffraction method. To obtain the most comprehensive information about a solid, other techniques besides may be used to complement a model based on diffraction data. These techniques include scanning electron microscopy (SEM), wavelength-dispersive analysis of X-rays (WDX), energy-dispersive analysis of X-rays (EDX), extended X-ray atomic fine-structure analysis (EXAFS), transmission electron microscopy (TEM), high-resolution transmission electron microscopy (HRTEM), differential thermal analysis (DTA), and a number of other methods.

For diffraction experiments, three types of radiation with a wavelength λ of the order of magnitude of interatomic distances are used: X-rays, electrons, and neutrons. The shortest interatomic distances in solids are a few times 10^{-10} m. Therefore the non-SI unit the angstrom (1 Å $= 10^{-10}$ m) is often used in crystallography. In the case of electrons and neutrons, their energies have to be converted to de Broglie wavelengths:

$$\lambda = h/mv\,,$$
$$\lambda(\text{Å}) = 0.28/\sqrt{E(\text{eV})}\,.$$

Figure 1.3-16 compares the energies and wavelengths of the three types of radiation.

From wave optics, it is known that radiation of wavelength λ is diffracted by a grid of spacing d. If we take a 3-D crystal lattice as such a grid, we expect diffraction maxima to occur at angles 2θ, given by the Bragg equation (Fig. 1.3-17)

$$\lambda = 2d_{hkl}\sin\theta_{hkl}\,.$$

For the aperiodic (n-D periodic crystal) case, d_{hkl} has to be replaced by $d_{h_1h_2...h_i...h_n}$. To give a simple 3-D example, for the determination of the cell parameter a in the cubic case, the Bragg equation can be rewritten in the form

$$(Q/2\pi)^2 = 4\sin^2\theta_{hkl}/\lambda^2 = (h^2+k^2+l^2)/a^2\,.$$

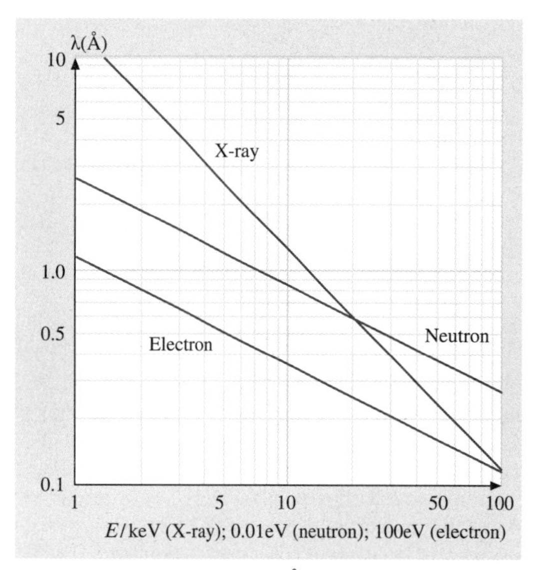

Fig. 1.3-16 Wavelengths λ in Å and particle energies E for X-ray photons (energies in keV), neutrons (energies in 0.01 eV), and electrons (energies in 100 eV)

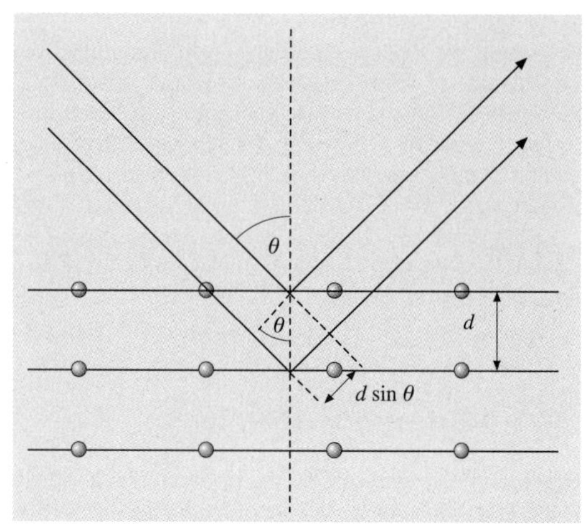

Fig. 1.3-17 Geometrical derivation of the Bragg equation $n\lambda = 2d\sin\theta$. n can be set to 1 when it is included in a higher-order hkl

Thus the crystal lattice is determined by a set of θ_{hkl}. In the case of X-rays and neutrons, information about the atomic structure is contained in the set of diffraction intensities I_{hkl}. Here we have $I_{hkl} = F_{hkl}^2$ where F_{hkl} are the structure factors.

To reconstruct the matter distribution $\rho(xyz)$ inside a unit cell of volume V, the crystallographic phase problem has to be solved. Once the phase factor ϕ for each hkl is known, the crystal structure is solved.

$$\rho(xyz) = 1/V \times \sum_{\text{all } h,k,l}\sum\sum |F|\cos[2\pi(hx+ky+lz)-\phi]\,.$$

Non-Bragg diffraction intensities $I(Q)$ and therefore a normalized structure function $S(Q)$ can be obtained, for example, from an X-ray or neutron powder diffractogram. The sine Fourier transform of $S(Q)$ yields a normalized radial atomic pair distribution function $G(r)$:

$$G(r) = (2/\pi)\int_0^\infty Q[S(Q)-1]\sin(Qr)\,dQ\,.$$

For measurements at high Q, the 1-D function $G(r)$ contains detailed information about the local structure. This function therefore resolves, for example, disorder or vacancy distributions in a material. The method can be applied to 3-D diffuse scattering distributions as well and thus can include angular information with respect to $\boldsymbol{r}$.

X-rays

X-rays can be produced in the laboratory using a conventional X-ray tube. Depending on the anode material, wavelengths λ from 0.56 Å (Ag Kα) to 2.29 Å (Cr Kα) can be generated. Filtered or monochromatized radiation is usually used to collect diffraction data, from either single crystals or polycrystalline fine powders. A continuous X-ray spectrum, obtained from a tungsten anode, for example, is used to obtain Laue images to check the quality, orientation, and symmetry of single crystals.

X-rays with a higher intensity, a tunable energy, a narrower distribution, and higher brilliance are provided by synchrotron radiation facilities.

X-rays interact with the electrons in a structure and therefore provide information about the electron density distribution – mainly about the electrons near the atomic cores.

Neutron Diffraction

Neutrons, generated in a nuclear reactor, are useful for complementing X-ray diffraction information. They interact with the atomic nuclei, and with the magnetic moments of unpaired electrons if they are present in a structure. Hydrogen atoms, which are difficult to locate using X-rays (the contain one electron, if at all, near the proton), give a far better contrast in neutron diffraction experiments. The exact positions of atomic nuclei permit "X minus N" structure determinations, so that the location of valence electrons can be made observable. Furthermore, the magnetic structure of a material can be determined.

Electron Diffraction

The third type of radiation which can be used for diffraction purposes is an electron beam; this is usually done in combination with TEM or HRTEM. Because electrons have only a short penetration distance – electrons, being charged particles, interact strongly with the material – electron diffraction is mainly used for thin crystallites, surfaces, and thin films. In the TEM mode, domains and other features on the nanometer scale are visible. Nevertheless, crystallographic parameters such as unit cell dimensions, and symmetry and space group information can be obtained from selected areas.

In some cases, information about, for example, stacking faults or superstructures obtained from an electron diffraction experiment may lead to a revised, detailed crystal structure model that is "truer" than the model which was originally deduced from X-ray diffraction data. If only small crystals of a material are available, crystallographic models obtained from unit cell and symmetry information can be simulated and then adapted to fit HRTEM results.

The descriptions above provide the equipment needed to understand the structure of solid matter on the atomic scale. The concepts of crystallography, the technical terms, and the language used in this framework have been presented. The complementarities of the various experimental methods used to extract coherent, comprehensive information from a sample of material have been outlined. The "rudiments" presented here, however, should be understood only as a first step into the fascinating field of the atomic structure of condensed matter.

References

3.1 L. V. Azaroff: *Elements of X-Ray Crystallography* (McGraw-Hill, New York 1968)

3.2 J. Pickworth Glusker, K. N. Trueblood: *Crystal Structure Analysis – A Primer* (Oxford Univ. Press, Oxford 1985)

3.3 E. R. Wölfel: *Theorie und Praxis der Röntgenstrukturanalyse* (Vieweg, Braunschweig 1987)

3.4 W. Kleber, H.-J. Bautsch, J. Bohm: *Einführung in die Kristallographie* (Verlag Technik, Berlin 1998)

3.5 C. Giacovazzo (Ed.): *Fundamentals of Crystallography*, IUCr Texts on Crystallography (Oxford Univ. Press., Oxford 1992)

3.6 C. Janot: *Quasicrystals – A Primer* (Oxford Univ. Press, Oxford 1992)

3.7 S. J. L. Billinge, T. Egami: *Underneath the Bragg Peaks: Structural Analysis of Complex Materials* (Elsevier, Amsterdam 2003)

3.8 T. Hahn (Ed.): *International Tables for Crystallography*, Vol. A (Kluwer, Dordrecht 1992)

3.9 A. V. Shubnikov, N. V. Belov: *Colored Symmetry* (Pergamon Press, Oxford 1964)

Part 2

The Eleme

Part 2 The Elements

2.1. The Elements

This section provides tables of the physical and physicochemical properties of the elements. Emphasis is given to properties of the elements in the condensed state. The tables are structured according to the Periodic Table of the elements. Most of the tables deal with the properties of elements of one particular group (column) of the Periodic Table. Only the elements of the first period (hydrogen and helium), the lanthanides, and the actinides are arranged according to the periods (rows) of the Periodic Table. This synoptic representation is intended to provide an immediate overview of the trends in the data for chemically related elements.

2.1.1 Introduction

2.1.1.1 How to Use This Section

To find properties of a specific element or group of elements, start from one of Tables 2.1-1 – 2.1-5 and proceed in one of the following ways:

1. If you know the *name* of the element, refer to Table 2.1-1, where an alphabetical list of the elements is given, together with the numbers of the pages where the properties of these elements will be found.
2. If you know the *chemical symbol* of the element, refer to Table 2.1-2, where an alphabetical list of element symbols is given, together with the numbers of the pages where the properties of the corresponding elements will be found.
3. If you know the *atomic number Z* of the element, refer to Table 2.1-3, where a list of the elements in order of atomic number is given, together with the numbers of the pages where the properties of these elements will be found.
4. If you know the *group of the Periodic Table* that contains the element of interest, refer to Table 2.1-4, which gives the numbers of the pages where the properties of the elements of each group will be found.
5. If you wish to look up the element in the *Periodic Table*, refer to Table 2.1-5, where the element symbol and the atomic number will be found. Then use Table 2.1-2 or Table 2.1-3 to find the numbers of the pages where the properties of the element of interest are tabulated.
6. Alternatively you can also find the name and the chemical symbol of the element you are looking for in the alphabetic index at the end of the volume. The index again will give you the first number of those pages on which

you can find the properties of the element described.

The data-tables corresponding to the Periodes and Groups of the Periodic Table are subdivided in the following way:

A. Atomic, ionic, and molecular properties.
B. Materials data:
   (a) Crystallographic properties.
   (b) Mechanical properties.
   (c) Thermal and thermodynamic properties.
   (d) Electronic, electromagnetic, and optical properties.
C. Allotropic and high-pressure modifications.
D. Ionic radii.

2.1.2 Description of Properties Tabulated

2.1.2.1 Parts A of the Tables

The properties tabulated in parts A of the tables concern the atomic, ionic, and molecular properties of the elements:

- The relative atomic mass, or atomic weight, A.
- The abundance in the lithosphere and in the sea.
- The atomic radius: the radius r_{cov} for single covalent bonding (after Pauling), the radius r_{met} for metallic bonding with a coordination number of 12 (after Pauling), the radius r_{vdW} for van der Waals bonding (after Bondi), and, for some elements, the radius r_{os} of the outer-shell orbital are given.
- The completely and partially occupied electron shells in the atom.
- The symbol for the electronic ground state.
- The electronic configuration.
- The oxidation states.
- The electron affinity.
- The electronegativity X_A (after Allred and Rochow).
- The first, second, third, and fourth ionization energies and the standard electrode potential E^0.
- The internuclear distance in the molecule.
- The dissociation energy of the molecule.

2.1.2.2 Parts B of the Tables

Parts B of the tables contain data on the macroscopic properties of the elements. Most of the data concern the condensed phases. If not indicated otherwise, the data in this section apply to the standard state of the element, that is, they are valid at standard temperature and pressure (STP, i. e. $T = 298.15$ K and $p = 100$ kPa $= 1$ bar). For those elements which are stable in the gas phase at STP, data are given for the macroscopic properties in the gas phase.

The quantities describing the physical and physicochemical properties of materials can be divided into two classes. The first class contains all those quantities which are not directly connected with external (generalized) forces, these quantities have well-defined values even in the absence of external forces. Some examples are the electronic ground-state configuration of the atom, the coordination number in the crystallized state and the surface tension in the liquid state. The second class contains those quantities which describe the response of the material to externally applied (generalized) forces F. Such a force might be a mechanical stress field, an electric or magnetic field, a field gradient, or a temperature gradient. The response of the material to the external force might be observed via a suitable observable O, such as a mechanical strain, an electric current density, a dielectric polarization, a magnetization, or a heat current density. Assuming homogeneous conditions, the dependence of the observable O on the force F can be used to define material-specific parameters χ, which are also called physical properties of the material. Some examples are the elastic moduli or compliance constants, the electrical conductivity, the dielectric constant, the magnetic susceptibility, and the thermal conductivity.

In the *linear-response* regime, that is, under weak external forces F, these parameters χ are considered as being independent of the strength of the forces. The dependence of an observable O on a force F is then the simple proportionality

$$O = \chi F \, . \tag{1.1}$$

For strong external fields, the dependence of the response on the strength of the forces can be expressed by a power expansion in the forces, which then – in addition to the linear parameters χ – defines *nonlinear field-dependent* materials properties $\chi^{(nl)}(F)$, where

$$\chi^{(nl)}(F) = \chi + \chi^{(1)} F + \ldots \, . \tag{1.2}$$

In general, the class of materials properties that describe the response to externally applied forces have *tensor* character. The *rank* of the property tensor χ depends on the rank of the external force F and that of the observable O considered. In the case of Ohm's law, $j = \sigma E$, in which the current density j and the electric field strength E are vectors, the conductivity tensor σ is of rank 2; in the case of the generalized Hooke's law, $\boldsymbol{\varepsilon} = \boldsymbol{s}\boldsymbol{\sigma}$, the strain tensor $\boldsymbol{\varepsilon}$ and the stress tensor $\boldsymbol{\sigma}$ both are of rank 2, so that the elastic compliance tensor $\boldsymbol{s}$ is of rank 4. A vector can be considered as a tensor of rank 1, and a scalar, correspondingly, as a tensor of rank 0. A second-rank tensor, such as the electrical conductivity $\boldsymbol{\sigma}$, in general has nine components in three-dimensional space; a tensor of rank n in general has 3^n components in three-dimensional space. *Symmetry*, however, of both the underlying crystal lattice and the physical phenomenon (for example, action = reaction), may reduce the number of independent nonvanishing components in the tensor. The tensor components reflect the crystal symmetry by being invariant under those orthogonal transformations which are elements of the point group of the crystal. In cubic crystals, for example, physical properties described by tensors of rank 2 are characterized by only one nonvanishing tensor component. Therefore cubic crystals are isotropic with respect to their electrical conductivity, their heat conductivity, and their dielectric properties.

Subdivisions B(a) of the Tables

These parts deal with the *crystallographic properties*. Here you will find the crystal system and the Bravais lattice in which the element is stable in its standard state; the structure type in which the element crystallizes; the lattice constants $a, b, c, \alpha, \beta, \gamma$ (symmetry reduces the number of independent lattice constants); the space group; the Schoenflies symbol; the *Strukturbericht* type; the Pearson symbol; the number A of atoms per cell; the coordination number; and the shortest interatomic distance between atoms in the solid state and in the liquid state.

Basic concepts of crystallography are explained in Chapt. 1.3.

Subdivisions B(b) of the Tables

These parts cover the *mechanical properties*. At the top of the table, you will find the density of the material in the solid state (ϱ_s) and in the liquid state (ϱ_l), and the molar volume V_{mol} in the solid state. Here, one mole is the amount of substance which contains as many elementary particles (atoms or molecules) as there are atoms in 0.012 kg of the carbon isotope with a relative atomic mass of 12. This number of particles is called *Avogadro's number* and is approximately equal to 6.022×10^{23}. The next three rows present the viscosity η, the surface tension, and its temperature dependence, in the liquid state. The next properties are the coefficient of linear thermal expansion α and the sound velocity, both in the solid and in the liquid state. A number of quantities are tabulated for the presentation of the elastic properties. For isotropic materials, we list the volume compressibility $\kappa = -(1/V)(dV/dP)$, and in some cases also its reciprocal value, the bulk modulus (or compression modulus); the elastic modulus (or Young's modulus) E; the shear modulus G; and the Poisson number (or Poisson's ratio) μ. Hooke's law, which expresses the linear relation between the strain ε and the stress σ in terms of Young's modulus, reads $\sigma = E\varepsilon$. For monocrystalline materials, the components of the elastic compliance tensor $\boldsymbol{s}$ and the components of the elastic stiffness tensor $\boldsymbol{c}$ are given. The elastic compliance tensor $\boldsymbol{s}$ and the elastic stiffness tensor $\boldsymbol{c}$ are both defined by the generalized forms of Hooke's law, $\boldsymbol{\sigma} = \boldsymbol{c}\boldsymbol{\varepsilon}$ and $\boldsymbol{\varepsilon} = \boldsymbol{s}\boldsymbol{\sigma}$. At the end of the list, the tensile strength, the Vickers hardness, and the Mohs hardness are given for some elements.

Subdivisions B(c) of the Tables

The *thermal and thermodynamic properties* are tabulated in these subdivisions of the tables. The properties tabulated are:

- The thermal conductivity λ.
- The molar heat capacity at constant pressure, c_p.
- The standard entropy S^0, that is, the molar entropy of the element at 298.15 K and 100 kPa.
- The enthalpy difference $H_{298} - H_0$, that is, the difference between the molar enthalpies of the element at 298.15 K and at 0 K.
- The melting temperature T_m.
- The molar enthalpy change ΔH_m and molar entropy change ΔS_m at the melting temperature.
- The relative volume change $\Delta V_m = (V_l - V_s)/V_l$ on melting.
- The boiling temperature T_b.
- The molar enthalpy change ΔH_b of boiling, and, for some elements, the molar enthalpy of sublimation.

In addition, the critical temperature T_c, the critical pressure p_c, the critical density ϱ_c, the triple-point temperature T_{tr}, and the triple-point pressure p_{tr} are given for some elements. For the element helium, the table also contains data for the λ point, at which liquid helium passes from the normal-fluid phase helium I (above the λ point) to the superfluid phase helium II (below the λ point), for ^4He and ^3He.

Throughout Sect. 2.1, temperature is measured in units of kelvin (K), the unit of thermodynamic temperature. 1 K is defined as the fraction 1/273.16 of the thermodynamic temperature of the triple point of water. To convert data given in kelvin into degrees Celsius (°C), the following equation can be used:

$$T(°C) = (T(K) - 273.15\,K)(°C/K) .$$

This can be expressed in words as follows: the Celsius scale is shifted towards higher temperatures by 273.15 K relative to the kelvin scale, such that the temperature 273.15 K becomes 0 °C and the temperature 0 K becomes −273.15 °C. To convert data given in kelvin into degrees Fahrenheit (°F), the following equation can be used:

$$T(°F) = (9/5)(T(K) - 273.15\,K)(°F/K) + 32\,°F .$$

This can be expressed approximately in words as follows: the Fahrenheit scale is shifted relative to the kelvin scale and also differs by a scaling so that its degrees are smaller than those of the kelvin scale by nearly a factor of 2.

Subdivisions B(d) of the Tables

These subdivisions of the tables present data on the *electronic, electromagnetic, and optical properties* of the elements. Data are given for the following:

- The electrical resistivity ρ_s in the solid state, and its temperature and pressure dependence.
- The electrical resistivity ρ_l in the liquid state, and the resistivity ratio ρ_l/ρ_s at the melting temperature.
- The critical temperature T_{cr} and critical field strength H_{cr} for superconductivity.
- The electronic band gap ΔE.
- The Hall coefficient R, together with the range of magnetic field strength B over which it was measured.
- The thermoelectric coefficient.
- The electronic work function.
- The thermal work function.
- The intrinsic charge carrier concentration.
- The electron and hole mobilities.
- The static dielectric constant ε of the element in the solid state, and in some cases also in the liquid state.
- The molar magnetic susceptibility χ_{mol} and the mass magnetic susceptibility χ_{mass} of the element in the solid state, and in some cases also in the liquid state. The susceptibilities are given in the definitions of both the SI system and the cgs system (see below).
- The refractive index n in the solid and liquid states.

The magnetic susceptibility is the parameter that describes the response of the material to an externally applied magnetic field H, as measured by the observable magnetization M, in the linear regime, via $M = \chi H$. Three different forms of the term "magnetization" are in use, depending on the specific application: first, the volume magnetization M_{vol}, equal to the magnetic dipole moment divided by the volume of the sample; second, the molar magnetization, or magnetization related to the number of particles, M_{mol}, equal to the magnetic dipole moment divided by the number of particles measures in moles; and third, the mass magnetization M_{mass}, equal to the magnetic dipole moment divided by the mass of the sample. Correspondingly, there are three different magnetic susceptibilities. The *volume susceptibility* χ_{vol} is a dimensionless number because in this case M and H are both measured in the same units, namely A/m in the SI system and gauss in the cgs system. The dimensionless character of χ_{vol} might be the reason why, in physics textbooks, mostly only this susceptibility is mentioned. The other two susceptibilities, the *molar susceptibility* χ_{mol} and the *mass susceptibility* χ_{mass}, are more useful for practical applications. In both the SI system and the cgs system, the molar susceptibility is measured in units of cm^3/mol, and the mass susceptibility is measured in units of cm^3/g. In this Handbook, data are given for the molar and mass susceptibilities.

Although susceptibilities have the same dimensions in the SI and cgs systems, the numerical values in the cgs system are smaller than those in the SI system by a factor of 4π. This is due to the different definitions of the quantities dipole moment and magnetization in the two systems. The difference can be seen most clearly in the general relations between the magnetization M and the field strengths B and H in the two systems. In the SI system, this relation reads $B = \mu_0(H + M)$, whereas in the cgs system, it reads $B = H + 4\pi M$. Because of this difference, the magnetic-susceptibility data in Sect. 2.1.5 are given for both the SI and the cgs definitions.

2.1.2.3 Parts C of the Tables

Parts C of the tables present crystallographic data for allotropic and high-pressure modifications of the elements. The left-hand columns contain data for allotropic modifications that are stable at a pressure of 100 kPa over the temperature ranges indicated, and the right-hand columns contain data for modifications stable at higher pressures as indicated. The modifications stable at 100 kPa are denoted by Greek letters in front of the chemical symbol of the element (normally starting with α for the modification stable over the lowest temperature range), and the high-pressure modifications are denoted by Roman numerals after the chemical symbol. In these parts of the tables, "RT" stands for "room temperature", and "RTP" stands for "room temperature and standard pressure", i.e. 100 kPa.

2.1.2.4 Parts D of the Tables

Parts D of the tables contain data on *ionic radii* determined from crystal structures. The first row lists the elements, and the second row lists the positive and negative ions for which data are given. The remaining rows give the ionic radii of these ions for the most common coordination numbers.

2.1.3 Sources

Most of the data presented here have been taken from Landolt–Börnstein [1.1]. Additional data have been taken from the D'Ans-Lax series [1.2] and the *CRC Handbook of Chemistry and Physics* [1.3].

2.1.4 Tables of the Elements in Different Orders

Table 2.1-1 The elements ordered by their names

Element	Symbol	Atomic Number	Page	Element	Symbol	Atomic Number	Page	Element	Symbol	Atomic Number	Page
Actinium	Ac	89	84	Gold	Au	79	65	Praseodymium	Pr	59	142
Aluminium	Al	13	78	Hafnium	Hf	72	94	Promethium	Pm	61	142
Americium	Am	95	151	Hassium	Hs	108	131	Protactinium	Pa	91	151
Antimony	Sb	51	98	Helium	He	2	54	Radium	Ra	88	68
Argon	Ar	18	128	Holmium	Ho	67	142	Radon	Rn	86	128
Arsenic	As	33	98	Hydrogen	H	1	54	Rhenium	Re	75	124
Astatine	At	85	118	Indium	In	49	78	Rhodium	Rh	45	135
Barium	Ba	56	68	Iodine	I	53	118	Roentgenium	Rg	111	
Berkelium	Bk	97	151	Iridium	Ir	77	135	Rubidium	Rb	37	59
Beryllium	Be	4	68	Iron	Fe	26	131	Ruthenium	Ru	44	131
Bismuth	Bi	83	98	Krypton	Kr	36	128	Rutherfordium	Rf	104	94
Bohrium	Bh	107	124	Lanthanum	La	57	84	Samarium	Sm	62	142
Boron	B	5	78	Lawrencium	Lr	103	151	Scandium	Sc	21	84
Bromine	Br	35	118	Lead	Pb	82	88	Seaborgium	Sg	106	114
Cadmium	Cd	48	73	Lithium	Li	3	59	Selenium	Se	34	108
Calcium	Ca	20	68	Lutetium	Lu	71	142	Silicon	Si	14	88
Californium	Cf	98	151	Magnesium	Mg	12	68	Silver	Ag	47	65
Carbon	C	6	88	Manganese	Mn	25	124	Sodium	Na	11	59
Cerium	Ce	58	142	Meitnerium	Mt	109	135	Strontium	Sr	38	68
Cesium	Cs	55	59	Mendelevium	Md	101	151	Sulfur	S	16	108
Chlorine	Cl	17	118	Mercury	Hg	80	73	Tantalum	Ta	73	105
Chromium	Cr	24	114	Molybdenum	Mo	42	114	Technetium	Tc	43	124
Cobalt	Co	27	135	Neodymium	Nd	60	142	Tellurium	Te	52	108
Copper	Cu	29	65	Neon	Ne	10	128	Terbium	Tb	65	142
Curium	Cm	96	151	Neptunium	Np	93	151	Thallium	Tl	81	78
Darmstadtium	Ds	110	139	Nickel	Ni	28	139	Thorium	Th	90	151
Dubnium	Db	105	105	Niobium	Nb	41	105	Thulium	Tm	69	142
Dysprosium	Dy	66	142	Nitrogen	N	7	98	Tin	Sn	50	88
Einsteinium	Es	99	151	Nobelium	No	102	151	Titanium	Ti	22	94
Erbium	Er	68	142	Osmium	Os	76	131	Tungsten	W	74	114
Europium	Eu	63	142	Oxygen	O	8	108	Uranium	U	92	151

Table 2.1-1 The elements ordered by their names, cont.

Element	Symbol	Atomic Number	Page	Element	Symbol	Atomic Number	Page	Element	Symbol	Atomic Number	Page
Fermium	Fm	100	151	Palladium	Pd	46	139	Vanadium	V	23	105
Fluorine	F	9	118	Phosphorus	P	15	98	Xenon	Xe	54	128
Francium	Fr	87	59	Platinum	Pt	78	139	Ytterbium	Yb	70	142
Gadolinium	Gd	64	142	Plutonium	Pu	94	151	Yttrium	Y	39	84
Gallium	Ga	31	78	Polonium	Po	84	108	Zinc	Zn	30	73
Germanium	Ge	32	88	Potassium	K	19	59	Zirconium	Zr	40	94

[a] See Tungsten.

Table 2.1-2 The elements ordered by their chemical symbols

Element	Symbol	Atomic Number	Page	Element	Symbol	Atomic Number	Page	Element	Symbol	Atomic Number	Page
Actinium	Ac	89	84	Gadolinium	Gd	64	142	Polonium	Po	84	108
Silver	Ag	47	65	Germanium	Ge	32	88	Praseodymium	Pr	59	142
Aluminium	Al	13	78	Hydrogen	H	1	54	Platinum	Pt	78	139
Americium	Am	95	151	Helium	He	2	54	Plutonium	Pu	94	151
Argon	Ar	18	128	Mercury	Hg	80	73	Radium	Ra	88	68
Arsenic	As	33	98	Hafnium	Hf	72	94	Rubidium	Rb	37	59
Astatine	At	85	118	Holmium	Ho	67	142	Rhenium	Re	75	124
Gold	Au	79	65	Hassium	Hs	108	131	Rutherfordium	Rf	104	94
Boron	B	5	78	Iodine	I	53	118	Roentgenium	Rg	111	
Barium	Ba	56	68	Indium	In	49	78	Rhodium	Rh	45	135
Beryllium	Be	4	68	Iridium	Ir	77	135	Radon	Rn	86	128
Bohrium	Bh	107	124	Potassium	K	19	59	Ruthenium	Ru	44	131
Bismuth	Bi	83	98	Krypton	Kr	36	128	Sulfur	S	16	108
Berkelium	Bk	97	151	Lanthanum	La	57	84	Antimony	Sb	51	98
Bromine	Br	35	118	Lithium	Li	3	59	Scandium	Sc	21	84
Carbon	C	6	88	Lawrencium	Lr	103	151	Selenium	Se	34	108
Calcium	Ca	20	68	Lutetium	Lu	71	142	Seaborgium	Sg	106	114
Cadmium	Cd	48	73	Mendelevium	Md	101	151	Silicon	Si	14	88
Cerium	Ce	58	142	Magnesium	Mg	12	68	Samarium	Sm	62	142
Californium	Cf	98	151	Manganese	Mn	25	124	Tin	Sn	50	88
Chlorine	Cl	17	118	Molybdenum	Mo	42	114	Strontium	Sr	38	68
Curium	Cm	96	151	Meitnerium	Mt	109	135	Tantalum	Ta	73	105
Cobalt	Co	27	135	Nitrogen	N	7	98	Terbium	Tb	65	142
Chromium	Cr	24	114	Sodium	Na	11	59	Technetium	Tc	43	124
Cesium	Cs	55	59	Niobium	Nb	41	105	Tellurium	Te	52	108
Copper	Cu	29	65	Neodymium	Nd	60	142	Thorium	Th	90	151
Dubnium	Db	105	105	Neon	Ne	10	128	Titanium	Ti	22	94
Darmstadtium	Ds	110	139	Nickel	Ni	28	139	Thallium	Tl	81	78
Dysprosium	Dy	66	142	Nobelium	No	102	151	Thulium	Tm	69	142
Erbium	Er	68	142	Neptunium	Np	93	151	Uranium	U	92	151
Einsteinium	Es	99	151	Oxygen	O	8	108	Vanadium	V	23	105
Europium	Eu	63	142	Osmium	Os	76	131	Tungsten	W	74	114
Fluorine	F	9	118	Phosphorus	P	15	98	Xenon	Xe	54	128
Iron	Fe	26	131	Protactinium	Pa	91	151	Yttrium	Y	39	84
Fermium	Fm	100	151	Lead	Pb	82	88	Ytterbium	Yb	70	142
Francium	Fr	87	59	Palladium	Pd	46	139	Zinc	Zn	30	73
Gallium	Ga	31	78	Promethium	Pm	61	142	Zirconium	Zr	40	94

Table 2.1-3 The elements ordered by their atomic numbers

Element	Symbol	Atomic Number	Page	Element	Symbol	Atomic Number	Page	Element	Symbol	Atomic Number	Page
Hydrogen	H	1	54	Strontium	Sr	38	68	Rhenium	Re	75	124
Helium	He	2	54	Yttrium	Y	39	84	Osmium	Os	76	131
Lithium	Li	3	59	Zirconium	Zr	40	94	Iridium	Ir	77	135
Beryllium	Be	4	68	Niobium	Nb	41	105	Platinum	Pt	78	139
Boron	B	5	78	Molybdenum	Mo	42	114	Gold	Au	79	65
Carbon	C	6	88	Technetium	Tc	43	124	Mercury	Hg	80	73
Nitrogen	N	7	98	Ruthenium	Ru	44	131	Thallium	Tl	81	78
Oxygen	O	8	108	Rhodium	Rh	45	135	Lead	Pb	82	88
Fluorine	F	9	118	Palladium	Pd	46	139	Bismuth	Bi	83	98
Neon	Ne	10	128	Silver	Ag	47	65	Polonium	Po	84	108
Sodium	Na	11	59	Cadmium	Cd	48	73	Astatine	At	85	118
Magnesium	Mg	12	68	Indium	In	49	78	Radon	Rn	86	128
Aluminium	Al	13	78	Tin	Sn	50	88	Francium	Fr	87	59
Silicon	Si	14	88	Antimony	Sb	51	98	Radium	Ra	88	68
Phosphorus	P	15	98	Tellurium	Te	52	108	Actinium	Ac	89	84
Sulfur	S	16	108	Iodine	I	53	118	Thorium	Th	90	151
Chlorine	Cl	17	118	Xenon	Xe	54	128	Protactinium	Pa	91	151
Argon	Ar	18	128	Cesium	Cs	55	59	Uranium	U	92	151
Potassium	K	19	59	Barium	Ba	56	68	Neptunium	Np	93	151
Calcium	Ca	20	68	Lanthanum	La	57	84	Plutonium	Pu	94	151
Scandium	Sc	21	84	Cerium	Ce	58	142	Americium	Am	95	151
Titanium	Ti	22	94	Praseodymium	Pr	59	142	Curium	Cm	96	151
Vanadium	V	23	105	Neodymium	Nd	60	142	Berkelium	Bk	97	151
Chromium	Cr	24	114	Promethium	Pm	61	142	Californium	Cf	98	151
Manganese	Mn	25	124	Samarium	Sm	62	142	Einsteinium	Es	99	151
Iron	Fe	26	131	Europium	Eu	63	142	Fermium	Fm	100	151
Cobalt	Co	27	135	Gadolinium	Gd	64	142	Mendelevium	Md	101	151
Nickel	Ni	28	139	Terbium	Tb	65	142	Nobelium	No	102	151
Copper	Cu	29	65	Dysprosium	Dy	66	142	Lawrencium	Lr	103	151
Zinc	Zn	30	73	Holmium	Ho	67	142	Rutherfordium	Rf	104	94
Gallium	Ga	31	78	Erbium	Er	68	142	Dubnium	Db	105	105
Germanium	Ge	32	88	Thulium	Tm	69	142	Seaborgium	Sg	106	114
Arsenic	As	33	98	Ytterbium	Yb	70	142	Bohrium	Bh	107	124
Selenium	Se	34	108	Lutetium	Lu	71	142	Hassium	Hs	108	131
Bromine	Br	35	118	Hafnium	Hf	72	94	Meitnerium	Mt	109	135
Krypton	Kr	36	128	Tantalum	Ta	73	105	Darmstadtium	Ds	110	139
Rubidium	Rb	37	59	Tungsten	W	74	114	Roentgenium	Rg	111	

Table 2.1-4 The elements ordered according to the Periodic Table

							Page
Elements of the first period							**54**
1 Hydrogen	1 Deuterium	1 Tritium	2 Helium-4	2 Helium-3			
Elements of Group IA							**59**
3 Lithium	11 Sodium	19 Potassium	37 Rubidium	55 Cesium	87 Francium		
Elements of Group IB							**65**
29 Copper	47 Silver	79 Gold	111 Roentgenium				
Elements of Group IIA							**68**
4 Beryllium	12 Magnesium	20 Calcium	38 Strontium	56 Barium	88 Radium		
Elements of Group IIB							**73**
30 Zinc	48 Cadmium	80 Mercury					
Elements of Group IIIA							**78**
5 Boron	13 Aluminium	31 Gallium	49 Indium	81 Thallium			
Elements of Group IIIB							**84**
21 Scandium	39 Yttrium	57 Lanthanum	89 Actinium				
Elements of Group IVA							**88**
6 Carbon	14 Silicon	32 Germanium	50 Tin	82 Lead			
Elements of Group IVB							**94**
22 Titanium	40 Zirconium	72 Hafnium	104 Rutherfordium				
Elements of Group VA							**98**
7 Nitrogen	15 Phosphorus	33 Arsenic	51 Antimony	83 Bismuth			
Elements of Group VB							**105**
23 Vanadium	41 Niobium	73 Tantalum	105 Dubnium				
Elements of Group VIA							**108**
8 Oxygen	16 Sulfur	34 Selenium	52 Tellurium	84 Polonium			
Elements of Group VIB							**114**
24 Chromium	42 Molybdenum	74 Tungsten	106 Seaborgium				
Elements of Group VIIA							**118**
9 Fluorine	17 Chlorine	35 Bromine	53 Iodine	85 Astatine			
Elements of Group VIIB							**124**
25 Manganese	43 Technetium	75 Rhenium	107 Bohrium				
Elements of Group VIIIA							**128**
10 Neon	18 Argon	36 Krypton	54 Xenon	86 Radon			
Elements of Group VIII(1)							**131**
26 Iron	44 Ruthenium	76 Osmium	108 Hassium				
Elements of Group VIII(2)							**135**
27 Cobalt	45 Rhodium	77 Iridium	109 Meitnerium				
Elements of Group VIII(3)							**139**
28 Nickel	46 Palladium	78 Platinum	110 Darmstadtium				
Lanthanides							**142**
58 Cerium	59 Praseodymium	60 Neodymium	61 Promethium	62 Samarium	63 Europium	64 Gadolinium	
65 Terbium	66 Dysprosium	67 Holmium	68 Erbium	69 Thulium	70 Ytterbium	71 Lutetium	
Actinides							**151**
90 Thorium	91 Protactinium	92 Uranium	93 Neptunium	94 Plutonium	95 Americium	96 Curium	
97 Berkelium	98 Californium	99 Einsteinium	100 Fermium	101 Mendelevium	102 Nobelium	103 Lawrencium	

Table 2.1-5 Periodic Table of the elements

Periodic Table of the Elements

18 VIIIA	IUPAC Notation CAS Notation
2 He	Atomic Number Element Symbol
(shaded)	Unstable Nuclei

Main Groups

1 IA	2 IIA	13 IIIA	14 IVA	15 VA	16 VIA	17 VIIA	18 VIIIA	Shells
1 H							2 He	K
3 Li	4 Be	5 B	6 C	7 N	8 O	9 F	10 Ne	K–L
11 Na	12 Mg	13 Al	14 Si	15 P	16 S	17 Cl	18 Ar	K–L–M
19 K	20 Ca	31 Ga	32 Ge	33 As	34 Se	35 Br	36 Kr	–L–M–N
37 Rb	38 Sr	49 In	50 Sn	51 Sb	52 Te	53 I	54 Xe	–M–N–O
55 Cs	56 Ba	81 Tl	82 Pb	83 Bi	84 Po	85 At	86 Rn	–N–O–P
87 Fr	88 Ra							–O–P–Q

Subgroups

3 IIIB	4 IVB	5 VB	6 VIB	7 VIIB	8 VIII (1)	9 VIII (2)	10 VIII (3)	11 IB	12 IIB	Shells
21 Sc	22 Ti	23 V	24 Cr	25 Mn	26 Fe	27 Co	28 Ni	29 Cu	30 Zn	–L–M–N
39 Y	40 Zr	41 Nb	42 Mo	43 Tc	44 Ru	45 Rh	46 Pd	47 Ag	48 Cd	–M–N–O
57 La	72 Hf	73 Ta	74 W	75 Re	76 Os	77 Ir	78 Pt	79 Au	80 Hg	–N–O–P
89 Ac	104 Rf	105 Db	106 Sg	107 Bh	108 Hs	109 Mt	110 Ds	111 Rg	112	–O–P–Q

Lanthanides (Shells –N–O–P)

58 Ce	59 Pr	60 Nd	61 Pm	62 Sm	63 Eu	64 Gd	65 Tb	66 Dy	67 Ho	68 Er	69 Tm	70 Yb	71 Lu

Actinides (Shells –O–P–Q)

90 Th	91 Pa	92 U	93 Np	94 Pu	95 Am	96 Cm	97 Bk	98 Cf	99 Es	100 Fm	101 Md	102 No	103 Lr

2.1.5 Data

2.1.5.1 Elements of the First Period

Table 2.1-6A Elements of the first period (hydrogen and helium). Atomic, ionic, and molecular properties

Element name	Hydrogen			Helium			
Special name	Hydrogen	Deuterium	Tritium	Helium 4	Helium 3		
Chemical symbol	H	^{2}H or D	^{3}H or T	^{4}He	^{3}He		
Atomic number Z	1	1	1	2	2		
						Units	Remarks
Characteristics			Radioactive				
Half-life			12.32			y	
Relative atomic mass A (atomic weight)	1.00794(7)	2.01408	3.01605	4.002602(2)			
Abundance in lithosphere	1400×10^{-6}	1.4×10^{-4}		4.2×10^{-7}			Mass ratio
Abundance in sea	110×10^{-3}						Mass ratio
Atomic radius r_{cov}	30					pm	Covalent radius
Atomic radius r_{met}				140		pm	Metallic radius, CN[a] = 12
Atomic radius r_{vdW}	140			140		pm	van der Waals radius
Electron shells	K	K	K	K	K		
Electronic ground state	$^2S_{1/2}$	$^2S_{1/2}$	$^2S_{1/2}$	$^1S_0$	$^1S_0$		
Electronic configuration	$1s^1$	$1s^1$	$1s^1$	$1s^2$	$1s^2$		
Oxidation states	1−, 1+						
Electron affinity	0.754			0.19		eV	
Electronegativity χ_A	2.20			3.20			Allred and Rochow
1st ionization energy	13.59844			24.58741		eV	
2nd ionization energy				54.41778		eV	
Standard electrode potential E^0	0.00000					V	Reaction $2H^+ + 2e^- = H_2$
Molecular form in gaseous state	H_2						
Internuclear distance in molecule	74.166					pm	
Dissociation energy	4.475					eV	Extrapolated to $T = 0$ K

[a] Coordination number.

Table 2.1-6B(a) Elements of the first period (hydrogen and helium). Crystallographic properties (for allotropic and high-pressure modifications, see Table 2.1-6C)

Element name	**Hydrogen**			**Helium**			
Special name	**Hydrogen**	**Deuterium**	**Tritium**	**Helium 4**	**Helium 3**		
Chemical symbol	**H**	**^{2}H or D**	**^{3}H or T**	**^{4}He**	**^{3}He**		
Atomic number *Z*	**1**	**1**	**1**	**2**	**2**		
						Units	**Remarks**
State	H_2 at 4.2 K			At 1.5 K			
Crystal system, Bravais lattice	hex			hex			
Structure type	Mg			Mg			
Lattice constant *a*	377.1			357.7		pm	
Lattice constant *c*	615.6			584.2			
Space group	$P6_3/mmc$			$P6_3/mmc$			
Schoenflies symbol	D_{6h}^4			D_{6h}^4			
Strukturbericht type	A3			AQ3			
Pearson symbol	hP2			hP2			
Number *A* of atoms per cell	2			2			
Coordination number	12			12			
Shortest interatomic distance, solid				357		pm	

Table 2.1-6B(b) Elements of the fist period (hydrogen and helium). Mechanical properties

Element name	Hydrogen			Helium			
Special name	Hydrogen	Deuterium	Tritium	Helium 4	Helium 3		
Chemical symbol	H	^{2}H or D	^{3}H or T	^{4}He	^{3}He		
Atomic number Z	1	1	1	2	2		
						Units	Remarks
State	H_2	D_2					
Density ϱ, liquid		0.162	0.260			g/cm^3	Near T_m
Density ϱ, gas	0.0899×10^{-3}	0.0032 [a]	0.0031 [a]	0.1785×10^{-3}		g/cm^3	At 273 K [b]
Molar volume V_{mol}, gas	13.26			32.07		cm^3/mol	
Viscosity η, gas	8.67					mPa s	At 293 K, 101 kPa
Sound velocity, gas	1237			969 (STP)		m/s	STP
Sound velocity, liquid	1340					m/s	At T_m, 12 MHz
Elastic modulus E	0.2			0.1		GPa	Solid state, estimated
Elastic compliance s_{11}	2930 (4.2 K)	1400 (4.2 K)		$+226 \times 10^3$ (1.6 K)	$+202 \times 10^3$ (0.4 K)	1/TPa	
Elastic compliance s_{33}	2000 (4.2 K)	995 (4.2 K)					
Elastic compliance s_{44}	9090 (4.2 K)	4350 (4.2 K)		$+46 \times 10^3$ (1.6 K)	$+108 \times 10^3$ (0.4 K)	1/TPa	
Elastic compliance s_{12}	−1240 (4.2 K)	−490 (4.2 K)		-107×10^3 (1.6 K)	-91.6×10^3 (0.4 K)	1/TPa	
Elastic compliance s_{13}	−166 (4.2 K)	−81 (4.2 K)					
Elastic stiffness c_{11}	0.042 (4.2 K)	0.82 (4.2 K)		$+31.1 \times 10^{-3}$ (1.6 K)	$+20.0 \times 10^{-3}$ (0.4 K)	GPa	
Elastic stiffness c_{33}	0.51 (4.2 K)	1.02 (4.2 K)					
Elastic stiffness c_{44}	0.11 (4.2 K)	0.23 (4.2 K)		$+21.7 \times 10^{-3}$ (1.6 K)	$+9.2 \times 10^{-3}$ (0.4 K)	GPa	
Elastic stiffness c_{12}	0.18 (4.2 K)	0.29 (4.2 K)		$+28.1 \times 10^{-3}$ (1.6 K)	$+16.4 \times 10^{-3}$ (0.4 K)	GPa	
Elastic stiffness c_{13}	0.05 (4.2 K)						
Solubility in water α_W	0.0178 [b]			0.0086			

[a] At 25 K.

[b] 101.3 kPa H_2 pressure.

Table 2.1-6B(c) Elements of the first period (hydrogen and helium). Thermal and thermodynamic properties

Element name	Hydrogen			Helium			
Special name	Hydrogen	Deuterium	Tritium	Helium 4	Helium 3		
Chemical symbol	H	^{2}H or D	^{3}H or T	^{4}He	^{3}He		
Atomic number Z	1	1	1	2	2		
						Units	Remarks
State	H_2 gas			He gas			
Thermal conductivity λ	0.171			143.0×10^{-3}		W/(m K)	At 300 K
Molar heat capacity c_p	14.418			20.786		J/(mol K)	At 298 K
Standard entropy S^0	130.680			126.152		J/(mol K)	At 298 K and 100 kPa
Enthalpy difference $H_{298} - H_0$	8.4684			6.1965		kJ/mol	At 298 K
Melting temperature T_m	13.81	18.65	20.65	0.95[a]		K	
Enthalpy change ΔH_m	0.12			0.021		kJ/mol	
Boiling temperature T_b	20.30	23.65	25.05	4.215	3.2	K	
Enthalpy change ΔH_b	0.46			0.082		kJ/mol	
Sublimation enthalpy	0.376			0.060		kJ/mol	At 0 K
Critical temperature T_c	33.2	38.55	39.95	5.23		K	
Critical pressure p_c	1.297	1.645		0.229		MPa	
Critical density ϱ_c	0.03102			69.3×10^{-3}		g/cm^3	
Triple-point temperature T_{tr}	14.0			[b]		K	
Triple-point pressure p_{tr}	7.2					kPa	
λ point				2.184	0.0027	K	

[a] At 2.6 MPa.

[b] He does not have a triple point.

Table 2.1-6B(d) Elements of the first period (hydrogen and helium). Electronic, electromagnetic, and optical properties

Element name	Hydrogen			Helium			
Special name	Hydrogen	Deuterium	Tritium	Helium 4	Helium 3		
Chemical symbol	H	^{2}H or D	^{3}H or T	^{4}He	^{3}He		
Atomic number Z	1	1	1	2	2		
						Units	**Remarks**
State	H_2 gas			He gas			
Characteristics	Light gas			Noble gas			
Dielectric constant $(\varepsilon - 1)$, gas	$+264 \times 10^{-6}$			$+68 \times 10^{-6}$			At 273 K
Dielectric constant ε, liquid	1.225			1.048 (3.15 K)			At 20.30 K
Dielectric constant ε, solid							
Molar magnetic susceptibility χ_{mol}, gas (SI system)	-50.1×10^{-6}			-25.4×10^{-6}		cm^3/mol	At 295 K
Molar magnetic susceptibility χ_{mol}, gas (cgs system)	-3.99×10^{-6}			-2.02×10^{-6}		cm^3/mol	At 295 K
Mass magnetic susceptibility χ_{mass}, liquid (SI system)	-68.4×10^{-6}					cm^3/mol	At 20.3 K
Mass magnetic susceptibility χ_{mass}, liquid (cgs system)	-5.44×10^{-6}					cm^3/mol	At 20.3 K
Mass magnetic susceptibility χ_{mass}, gas (SI system)	-25×10^{-6}			-5.9×10^{-6}		cm^3/g	At 295 K
Refractive index $(n - 1)$, gas	132×10^{-6}			36×10^{-6}			At 273.15 K, 101×10^5 Pa, $\lambda = 589.3$ nm
Refractive index n, liquid	1.112			1.026 (3.7 K)			$\lambda = 589.3$ nm

Table 2.1-6C Elements of the first period (hydrogen and helium). Allotropic and high-pressure modifications

Element	Hydrogen		Helium			
Modification	α-H (H_2)	β-H (H_2)	α-He	β-He	γ-He	
						Units
Crystal system, Bravais lattice	cub, fc	hex, cp	hex, cp	cub, fc	cub, bc	
Structure type	Cu	Mg	Mg	Cu	W	
Lattice constant a	533.4	377.1	357.7	4240	4110	pm
Lattice constant c		615.6	584.2			pm
Space group	$Fm\bar{3}m$	$P6_3/mmc$	$P6_3/mmc$	$Fm\bar{3}m$	$Im\bar{3}m$	
Schoenflies symbol	O_h^5	D_{6h}^4	D_{6h}^4	O_h^5	O_h^9	
Strukturbericht type	A1	A3	A3	A1	A2	
Pearson symbol	cF4	hP2	hP2	cF4	cI2	
Number A of atoms per cell	4	2	2	4	2	
Coordination number	12	12	12	12	8	
Shortest interatomic distance, solid			357	300	356	pm
Range of stability	< 1.25 K	< 13.81 K	< 0.95 K	1.6 K; 0.125 GPa	1.73 K; 0.03 GPa	

2.1.5.2 Elements of the Main Groups and Subgroup I to IV

Table 2.1-7A Elements of Group IA (CAS notation), or Group 1 (new IUPAC notation). Atomic, ionic, and molecular properties (see Table 2.1-7D for ionic radii)

Element name	Lithium	Sodium	Potassium	Rubidium	Cesium	Francium		
Chemical symbol	Li	Na	K	Rb	Cs	Fr		
Atomic number Z	3	11	19	37	55	87		
							Units	Remarks
Characteristics						Radioactive		
Relative atomic mass A (atomic weight)	6.941(2)	22.989770(2)	39.0983(1)	85.4678(3)	132.90545(2)	[223]		
Abundance in lithosphere	65×10^{-6}	$28\,300 \times 10^{-6}$	$25\,900 \times 10^{-6}$	280×10^{-6}	3.2×10^{-6}			Mass ratio
Abundance in sea	0.18×10^{-6}	$10\,770 \times 10^{-6}$	380×10^{-6}	0.12×10^{-6}	4×10^{-10}			Mass ratio
Atomic radius r_{cov}	123	157	203	216	253		pm	Covalent radius
Atomic radius r_{met}	156	192	238	250	272		pm	Metallic radius, CN = 12
Atomic radius r_{os}	158.6	171.3	216.2	228.7	251.8	244.7	pm	Outer-shell orbital radius
Atomic radius r_{vdW}	180	230	280	244	262		pm	van der Waals radius
Electron shells	KL	KLM	–LMN	–MNO	–NOP	–OPQ		
Electronic ground state	$^2S_{1/2}$	$^2S_{1/2}$	$^2S_{1/2}$	$^2S_{1/2}$	$^2S_{1/2}$	$^2S_{1/2}$		
Electronic configuration	[He]$2s^1$	[Ne]$3s^1$	[Ar]$4s^1$	[Kr]$5s^1$	[Xe]$6s^1$	[Rn]$7s^1$		
Oxidation states	1+	1+	1+	1+	1+	1+		
Electron affinity	0.618	0.548	0.501	0.486	0.472	0.46	eV	
Electronegativity χ_A	0.97	1.01	0.91	0.89	0.86	(0.86)		Allred and Rochow
1st ionization energy	5.39172	5.13908	4.34066	4.17713	3.89390	4.0727	eV	
2nd ionization energy	75.64018	47.2864	31.63	27.285	23.15745		eV	
3rd ionization energy	122.45429	71.6200	45.806	40			eV	
4th ionization energy		98.91	60.91	52.6			eV	
Standard electrode potential E^0	−3.040	−2.71	−2.931	−2.98	−2.92		V	Reaction type $Li^+ + e^- = Li$

Table 2.1-7B(a) Elements of Group IA (CAS notation), or Group 1 (new IUPAC notation). Crystallographic properties (see Table 2.1-7C for allotropic and high-pressure modifications)

Element name	**Lithium**	**Sodium**	**Potassium**	**Rubidium**	**Cesium**	**Francium**		
Chemical symbol	**Li**	**Na**	**K**	**Rb**	**Cs**	**Fr**		
Atomic number *Z*	**3**	**11**	**19**	**37**	**55**	**87**		
							Units	**Remarks**
Modification	β-Li	β-Na				Cs-I		
Crystal system, Bravais lattice	cub, bc	cub, bc	cub, bc	cub, bc	cub, bc			
Structure type	W	W	W	W	W			
Lattice constant *a*	350.93	420.96	532.1	570.3	614.1		pm	
Space group	$Im\bar{3}m$	$Im\bar{3}m$	$Im\bar{3}m$	$Im\bar{3}m$	$Im\bar{3}m$			
Schoenflies symbol	O_h^9	O_h^9	O_h^9	O_h^9	O_h^9			
Strukturbericht type	A2	A2	A2	A2	A2			
Pearson symbol	cI2	cI2	cI2	cI2	cI2			
Number *A* of atoms per cell	2	2	2	2	2			
Coordination number	8	8	8	8	8			
Shortest interatomic distance, solid	303	371	462	487	524		pm	
Shortest interatomic distance, liquid				497 (313 K)			pm	

Table 2.1–7B(b) Elements of Group IA (CAS notation), or Group 1 (new IUPAC notation). Mechanical properties

Element name	Lithium	Sodium	Potassium	Rubidium	Cesium	Francium		
Chemical symbol	Li	Na	K	Rb	Cs	Fr		
Atomic number Z	3	11	19	37	55	87		
							Units	Remarks
Density ϱ, solid	0.532	0.970	0.862	1.532	1.87	2.410	g/cm^3	
Density ϱ, liquid	0.508 (453 K)	0.927	0.827	1.470	1.847		g/cm^3	Near T_m
Molar volume V_{mol}	13.00	23.68	45.36	55.79	70.96	9.25	cm^3/mol	
Viscosity η, liquid	0.566 (473 K)	0.565 (416 K)	0.64	0.52			mPa s	
Surface tension, liquid	0.396	0.193	0.116	0.092	0.060		N/m	
Temperature coefficient		-0.05×10^{-3}	-0.06×10^{-3}		-0.046×10^{-3}		N/(m K)	
Coefficient of linear thermal expansion α	56×10^{-6}	70.6×10^{-6}	83×10^{-6}	90×10^{-6}	97×10^{-6}		1/K	At 298 K
Sound velocity, liquid		2395 (371 K)		1260	967		m/s	At T_m, At 12 MHz
Sound velocity, solid, transverse	2820	1620	1230	770	590		m/s	
Sound velocity, solid, longitudinal	6030	3310	2600	1430	1090		m/s	
Compressibility κ	8.93×10^{-5}	13.4×10^{-5}	23.7×10^{-5}	33.0×10^{-5}	0.75×10^{-5}		1/MPa	Volume compressibility
Elastic modulus E	11.5	6.80	3.52	2.35	1.69		GPa	
Shear modulus G	4.24	2.94	1.28	0.91	0.65		GPa	
Poisson number μ	0.36	0.34	0.35	0.29	0.30			
Elastic compliance s_{11}	315	549	1339	1330	1190 (280 K)		1/TPa	
Elastic compliance s_{44}	104	233	526	625	690 (280 K)		1/TPa	
Elastic compliance s_{12}	−144	−250	−620	−600	−450 (280 K)		1/TPa	
Elastic stiffness c_{11}	13.4	7.59	3.69	2.96	1.60 (280 K)		GPa	
Elastic stiffness c_{44}	9.6	4.30	1.90	1.60	1.44 (280 K)		GPa	
Elastic stiffness c_{12}	11.3	6.33	3.18	2.44	0.99 (280 K)		GPa	
Tensile strength	0.6						MPa	
Vickers hardness				0.37	0.15			
Mohs hardness	0.6							

Table 2.1–7B(c) Elements of Group IA (CAS notation), or Group 1 (new IUPAC notation). Thermal and thermodynamic properties

Element name	Lithium	Sodium	Potassium	Rubidium	Cesium	Francium		
Chemical symbol	**Li**	**Na**	**K**	**Rb**	**Cs**	**Fr**		
Atomic number Z	**3**	**11**	**19**	**37**	**55**	**87**		
							Units	**Remarks**
Thermal conductivity λ	84.7	141	102.4	58.2	35.9		W/(m K)	At 300 K
Molar heat capacity c_p	24.77	28.24	29.58	31.062	32.18		J/(mol K)	At 298 K
Standard entropy S^0	29.120	51.300	64.680	76.776	85.230	101.00	J/(mol K)	At 298 K and 100 kPa
Enthalpy difference $H_{298} - H_0$	4.6320	6.4600	7.0880	7.4890	7.7110	10.000	kJ/mol	
Melting temperature T_m	453.69	370.87	336.86	312.47	301.59	300	K	
Enthalpy change ΔH_m	3.0000	2.5970	2.3208	2.1924	2.0960		kJ/mol	
Entropy change ΔS_m	6.612	7.002	6.890	7.016	6.950		J/(mol K)	
Relative volume change ΔV_m	0.0151	0.027	0.0291	0.0228	0.0263			$(V_l - V_s)/V_l$ at T_m
Boiling temperature T_b	1620	1156	1040	970	947	950 (estimated)	K	
Enthalpy change ΔH_b	147.7	99.2	79.1	75.7	67.7		kJ/mol	
Critical temperature T_c		2500	2280		2010		K	
Critical pressure p_c		25.3	16.1		11.3		MPa	
Critical density ϱ_c		0.210	0.190		0.410		g/cm^3	

Table 2.1-7B(d) Elements of Group IA (CAS notation), or Group 1 (new IUPAC notation). Electronic, electromagnetic, and optical properties

Element name	Lithium	Sodium	Potassium	Rubidium	Cesium	Francium		
Chemical symbol	Li	Na	K	Rb	Cs	Fr		
Atomic number Z	3	11	19	37	55	87		
							Units	Remarks
Characteristics	Very reactive metal	Reactive metal	Soft, reactive metal	Soft, reactive metal	Alkali metal	Alkali metal		
Electrical resistivity ρ_s	85.5	42	61	116	188		nΩ m	Solid, at 293 K
Temperature coefficient	48.9×10^{-4}	54.6×10^{-4}	67.3×10^{-4}	63.7×10^{-4}	50.3×10^{-4}		1/K	Solid
Pressure coefficient	-2.1×10^{-9}	-38.3×10^{-9}	-69.7×10^{-9}	-62.9×10^{-9}	0.5×10^{-9}		1/hPa	Solid
Electrical resistivity ρ_l	240 at T_m		129.7	220	367		nΩ m	Liquid
Resistivity ratio at T_m	1.68	1.44	1.56	1.612	1.66			ρ_l/ρ_s at T_m
Hall coefficient R[a]	-1.70×10^{-10}	-2.1×10^{-10}	-4.2×10^{-10}	-5.92×10^{-10}	-7.8×10^{-10}		$m^3/(A\,s)$	At 300 K
Thermoelectric coefficient	14.37		12	−8.26	0.2		μV/K	
Electronic work function	2.28	2.75	2.30	2.05	1.94		V	
Thermal work function	2.39		2.15	2.13	1.87		V	
Molar magnetic susceptibility χ_{mol}, solid (SI)	178×10^{-6}	201×10^{-6}	261×10^{-6}	214×10^{-6}	364×10^{-6}		cm^3/mol	At 295 K
Molar magnetic susceptibility χ_{mol}, solid (cgs)	14.2×10^{-6}	16×10^{-6}	20.8×10^{-6}	17×10^{-6}	29×10^{-6}		cm^3/mol	At 295 K
Mass magnetic susceptibility χ_{mass}, solid (SI)	25.6×10^{-6}	8.8×10^{-6}	6.7×10^{-6}	2.49×10^{-6}	2.8×10^{-6}		cm^3/g	At 295 K
Mass magnetic susceptibility χ_{mass}, liquid (SI)					2.6×10^{-6}		cm^3/g	
Refractive index n, solid		4.22	0.024 (134 nm)					$\lambda = 589.3$ nm
Refractive index n, liquid		0.0045						$\lambda = 589.3$ nm

[a] $B = 1\text{–}3\ T$

Table 2.1-7C Elements of Group IA (CAS notation), or Group 1 (new IUPAC notation). Allotropic and high-pressure modifications

Element	Lithium			Sodium		Cesium			
Modification	α-Li	β-Li	γ-Li	α-Na	β-Na	Cs-I	Cs-II	Cs-III	
									Units
Crystal system, Bravais lattice	hex, cp	cub, bc	cub, fc	hex, cp	cub, bc	cub, bc	cub, fc	cub, fc	
Structure type	Mg	W	Cu	Mg	W	W	Cu	Cu	
Lattice constant *a*	311.1	309.3		376.7	420.96	614.1	598.4	580.0	pm
Lattice constant *c*	509.3			615.4					
Space group	$P6_3/mmc$	$Im\bar{3}m$	$Fm\bar{3}m$	$P6_3/mmc$	$Im\bar{3}m$	$Im\bar{3}m$	$Fm\bar{3}m$	$Fm\bar{3}m$	
Schoenflies symbol	D_{6h}^4	O_h^9	O_h^5	D_{6h}^4	O_h^9	O_h^9	O_h^5	O_h^5	
Strukturbericht type	A3	A2	A1	A3	A2	A2	A1	A1	
Pearson symbol	hP2	cI2	cF4	hP2	cI2	cI2	cF4	cF4	
Number *A* of atoms per cell	2	2	4	2	2	2	4	4	
Coordination number	6+6	8	12	12	8	8	12	12	
Shortest interatomic distance, solid	310	303	310	377	371	524	457	410	pm
Range of stability	< 72 K	RT		< 36 K	RT	RTP	> 2.37 GPa	> 4.22 GPa	

Table 2.1-7D Elements of Group IA (CAS notation), or Group 1 (new IUPAC notation). Ionic radii (determined from crystal structures)

Element	Lithium	Sodium	Potassium	Rubidium	Cesium	Francium	
Ion	Li^+	Na^+	K^+	Rb^+	Cs^+	Fr^+	
Coordination number							**Units**
4	59	99	137				pm
6	76	102	138	152	167	180	pm
8	92	118	151	161	174		pm
9		124					pm
10				166	181		pm
12		139	164	172	188		pm

Table 2.1-8A Elements of Group IB (CAS notation), or Group 11 (new IUPAC notation). Atomic, ionic, and molecular properties (see Table 2.1-8D for ionic radii)

Element name	**Copper**	**Silver**	**Gold**		
Chemical symbol	**Cu**	**Ag**	**Au**		
Atomic number Z	**29**	**47**	**79**		
				Units	**Remarks**
Relative atomic mass A_r (atomic weight)	63.546(3)	107.8682(2)	196.96655(2)		
Abundance in lithosphere	70×10^{-6}	2.0×10^{-6}	3×10^{-9}		Mass ratio
Abundance in sea	5×10^{-10}	4×10^{-11}	4×10^{-12}		Mass ratio
Atomic radius r_{cov}	117	134	134	pm	Covalent radius
Atomic radius r_{met}	128	144	144	pm	Metallic radius, CN = 12
Atomic radius r_{vdW}	140	170	170	pm	van der Waals radius
Electron shells	–LMN	–MNO	–NOP		
Electronic ground state	$^2S_{1/2}$	$^2S_{1/2}$	$^2S_{1/2}$		
Electronic configuration	$[Ar]3d^{10}4s^1$	$[Kr]4d^{10}5s^1$	$[Xe]4f^{14}5d^{10}6s^1$		
Oxidation states	2+, 1+	1+	3+, 1+		
Electron affinity	1.24	1.30	2.309	eV	
Electronegativity χ_A	1.75	1.42	(1.42)		Allred and Rochow
1st ionization energy	7.72638	7.57624	9.22567	eV	
2nd ionization energy	20.29240	21.49	20.5	eV	
3rd ionization energy	36.841	34.83		eV	
4th ionization energy	57.38			eV	
Standard electrode potential E^0	+0.521	+0.7996	+1.692	V	Reaction type $Cu^+ + e^- = Cu$
	+0.342			V	Reaction type $Cu^{2+} + 2e^- = Cu$
	+0.153			V	Reaction type $Cu^{2+} + e^- = Cu^+$
			+1.498	V	Reaction type $Au^{3+} + 3e^- = Au$
			+1.401	V	Reaction type $Au^{3+} + 2e^- = Au^+$

Table 2.1-8B(a) Elements of Group IB (CAS notation), or Group 11 (new IUPAC notation). Crystallographic properties

Element name	**Copper**	**Silver**	**Gold**		
Chemical symbol	**Cu**	**Ag**	**Au**		
Atomic number Z	**29**	**47**	**79**		
				Units	**Remarks**
Crystal system, Bravais lattice	cub, fc	cub, fc	cub, fc		
Structure type	Cu	Cu	Cu		
Lattice constant a	361.49	408.61	407.84	pm	At 298 K
Space group	$Fm\bar{3}m$	$Fm\bar{3}m$	$Fm\bar{3}m$		
Schoenflies symbol	O_h^5	O_h^5	O_h^5		
Strukturbericht type	A1	A1	A1		
Pearson symbol	cF4	cF4	cF4		
Number A of atoms per cell	4	4	4		
Coordination number	12	12	12		
Shortest interatomic distance, solid	255.6	288	288	pm	At 293 K
Shortest interatomic distance, liquid	257 (1363 K)			pm	

Table 2.1-8B(b) Elements of Group IB (CAS notation), or Group 11 (new IUPAC notation). Mechanical properties

Element name	Copper	Silver	Gold		
Chemical symbol	Cu	Ag	Au		
Atomic number Z	29	47	79		
				Units	Remarks
Density ϱ, solid	8.960	10.50	19.30	g/cm^3	
Density ϱ, liquid	8.000	9.345	17.300	g/cm^3	
Molar volume V_{mol}	7.09	10.27	10.19	cm^3/mol	
Viscosity η, liquid	3.36	3.62	5.38	mPa s	At T_m
Surface tension, liquid	1.300	0.923	1.128	N/m	
Temperature coefficient	-0.18×10^{-3}	-0.13×10^{-3}	-0.10×10^{-3}	N/(m K)	
Coefficient of linear thermal expansion α	16.5×10^{-6}	19.2×10^{-6}	14.16×10^{-6}	1/K	At 298 K
Sound velocity, solid, transverse	2300	1690	1190	m/s	
Sound velocity, solid, longitudinal	4760	3640	3280	m/s	
Compressibility κ	0.702×10^{-5}	0.95×10^{-5}	0.563×10^{-5}	1/MPa	Volume compressibility
Elastic modulus E	128	80.0	78.0	GPa	
Shear modulus G	46.8	29.5	26.0	GPa	
Poisson number μ	0.34	0.38	0.42		
Elastic compliance s_{11}	15.0	23.0	23.4	1/TPa	
Elastic compliance s_{44}	13.3	22.0	23.8	1/TPa	
Elastic compliance s_{12}	−6.3	−9.8	−10.8	1/TPa	
Elastic stiffness c_{11}	169	122	191	GPa	
Elastic stiffness c_{44}	75.3	45.5	42.2	GPa	
Elastic stiffness c_{12}	122	92	162	GPa	
Tensile strength	209	125	110	MPa	At 293 K
Vickers hardness	369	251	216		At 293 K

Table 2.1-8B(c) Elements of Group IB (CAS notation), or Group 11 (new IUPAC notation). Thermal and thermodynamic properties

Element name	Copper	Silver	Gold		
Chemical symbol	Cu	Ag	Au		
Atomic number Z	29	47	79		
				Units	Remarks
Thermal conductivity λ	1401	429	317	W/(m K)	At 300 K
Molar heat capacity c_p	24.443	25.36	25.42	J/(mol K)	At 298 K
Standard entropy S^0	33.150	42.551	47.488	J/(mol K)	At 298 K and 100 kPa
Enthalpy difference $H_{298} - H_0$	5.0040	5.7446	6.0166	kJ/mol	
Melting temperature T_m	1357.77	1234.93	1337.33	K	
Enthalpy change ΔH_m	13.263	11.297	12.552	kJ/mol	
Entropy change ΔS_m	9.768	9.148	9.386	J/(mol K)	
Relative volume change	+0.0415	0.038	0.051		$(V_l - V_s)/V_l$ at T_m
Boiling temperature T_b	2843	2436	3130	K	
Enthalpy change ΔH_b	300.7	250.6	334.4	kJ/mol	

Table 2.1-8B(d) Elements of Group IB (CAS notation), or Group 11 (new IUPAC notation). Electronic, electromagnetic, and optical properties. There is no Table 2.1-8C, because no allotropic or high-pressure modifications are known

Element name	Copper	Silver	Gold		
Chemical symbol	Cu	Ag	Au		
Atomic number Z	29	47	79		
				Units	Remarks
Characteristics	Metal	Noble metal	Noble metal		
Electrical resistivity ρ_s	16.78	147	20.5	nΩ m	At 293 K
Temerature coefficient	43.8×10^{-4}	43×10^{-4}	40.2×10^{-4}	1/K	
Pressure coefficient	-1.86×10^{-9}	-3.38×10^{-9}	-2.93×10^{-9}	1/hPa	
Electrical resistivity ρ_l	21.5		312	nΩ m	
Resistivity ratio	2.07		2.28		ρ_l/ρ_s at T_m
Hall coefficient R	-0.536×10^{-10}	-0.84×10^{-10}	-0.704×10^{-10}	$m^3/(A\,s)$	$B=0.3$–2.2 T, 298 K
Thermoelectric coefficient	1.72	1.42	1.72	μV/K	
Electronic work function	4.65	4.26	5.1	V	
Thermal work function	4.39	4.31	4.25	V	
Molar magnetic susceptibility chi_{mol}, solid (SI)	-68.6×10^{-6}	-245×10^{-6}	-352×10^{-6}	cm^3/mol	At 295 K
Molar magnetic susceptibility χ_{mol}, solid (cgs)	-5.46×10^{-6}	-19.5×10^{-6}	-28×10^{-6}	cm^3/mol	At 295 K
Mass magnetic susceptibility χ_{mass}, solid (SI)	-1.081×10^{-6}	-2.27×10^{-6}	-1.78×10^{-6}	cm^3/g	At 295 K
Mass magnetic susceptibility χ_{mass}, liquid (SI)	-1.2×10^{-6}	-2.83×10^{-6}	-2.16×10^{-6}	cm^3/g	

Table 2.1-8D Elements of Group IB (CAS notation), or Group 11 (new IUPAC notation). Ionic radii (determined from crystal structures)

Element	Copper		Silver		Gold		
Ion	Cu^+	Cu^{2+}	Ag^+	Ag^{2+}	Au^+	Au^{3+}	
Coordination number							Units
2	46						pm
4	60	57	100	79		64	pm
6	77	73	115	94	137	85	pm
8			128				pm

Table 2.1-9A Elements of Group IIA (CAS notation), or Group 2 (new IUPAC notation). Atomic, ionic, and molecular properties (see Table 2.1-9D for ionic radii)

Element name	Beryllium	Magnesium	Calcium	Strontium	Barium	Radium		
Chemical symbol	Be	Mg	Ca	Sr	Ba	Ra		
Atomic number Z	4	12	20	38	56	88		
							Units	Remarks
Characteristics						Radioactive		
Relative atomic mass A (atomic weight)	9.012182(3)	24.3050(6)	40.078(4)	87.62(1)	137.327(7)	[226]		
Abundance in lithosphere	6×10^{-6}	$20\,900\times10^{-6}$	$26\,300\times10^{-6}$	150×10^{-6}	430×10^{-6}			Mass ratio
Abundance in sea	6×10^{-13}	1290×10^{-6}	412×10^{-6}	8.0×10^{-6}	2×10^{-9}			Mass ratio
Atomic radius r_{cov}	89	136	174	191	198		pm	Covalent radius
Atomic radius r_{met}	112	160	197	215	224	230	pm	Metallic radius, CN = 12
Atomic radius r_{vdW}		170					pm	van der Waals radius
Electron shells	KL	KLM	–LMN	–MNO	–NOP	–OPQ		
Electronic ground state	$^1S_0$	$^1S_0$	$^1S_0$	$^1S_0$	$^1S_0$	$^1S_0$		
Electronic configuration	[He]2s^2	[Ne]2s^2	[Ar]4s^2	[Kr]5ss^2	[Xe]6s^2	[Rn]7s^2		
Oxidation states	2+	2+	2+	2+	2+	2+		
Electron affinity	Not stable	Not stable	0.0246	0.048	0.15		eV	
Electronegativity χ_A	1.47	1.23	1.04	0.99	0.97	(0.97)		Allred and Rochow
1st ionization energy	9.32263	7.64624	6.11316	5.69484	5.21170	5.27892	eV	
2nd ionization energy	18.21116	15.03528	11.87172	11.03013	10.00390	10.14716	eV	
3rd ionization energy	153.89661	80.1437	50.9131	42.89			eV	
4th ionization energy	217.71865	109.2655	67.27	57			eV	
Standard electrode potential E^0	−1.847	−2.372	−2.868	−2.89	−2.90		V	Reaction type $Be^{++} + 2e^- = Be$

Table 2.1-9B(a) Elements of Group IIA (CAS notation), or Group 2 (new IUPAC notation). Crystallographic properties (see Table 2.1-9C for allotropic and high-pressure modifications)

Element name	Beryllium	Magnesium	Calcium	Strontium	Barium	Radium		
Chemical symbol	Be	Mg	Ca	Sr	Ba	Ra		
Atomic number Z	4	12	20	38	56	88		
							Units	Remarks
Modification	α-Be		α-Ca	α-Sr				
Crystal system, Bravais lattice	hex	hex	cub, fc	cub, fc	cub, bc	cub, bc		
Structure type	Mg	Mg	Cu	Cu	W	W		
Lattice constant a	228.57	320.93	558.84	608.4	502.3	514.8	pm	
Lattice constant c	358.39	521.07					pm	
Space group	$P6_3/mmc$	$P6_3/mmc$	$Fm\bar{3}m$	$Fm\bar{3}m$	$Im\bar{3}m$	$Im\bar{3}m$		
Schoenflies symbol	D_{6h}^4	D_{6h}^4	O_h^5	O_h^5	O_h^9	O_h^9		
Strukturbericht type	A3	A3	A1	A1	A2	A2		
Pearson symbol	hP2	hP2	cF4	cF4	cI2	cI2		
Number A of atoms per cell	2	2	4	4	2	2		
Coordination number	12	6+6	12	12	8	8		
Shortest interatomic distance, solid	222	319	393	430	434	446	pm	

Table 2.1-9B(b) Elements of Group IIA (CAS notation), or Group 2 (new IUPAC notation). Mechanical properties

Element name	Beryllium	Magnesium	Calcium	Strontium	Barium	Radium		
Chemical symbol	Be	Mg	Ca	Sr	Ba	Ra		
Atomic number Z	4	12	20	38	56	88		
							Units	Remarks
Density ϱ, solid	1.85	1.74	1.55	2.60	3.50	5.000	g/cm^3	
Density ϱ, liquid	1.420 (1770 K)	1.590	1.365	2.375	3.325		g/cm^3	
Molar volume V_{mol}	4.88	13.98	25.86	34.50	38.21	45.2	cm^3/mol	
Viscosity η, liquid		1.23	1.06				mPa s	
Surface tension γ, liquid	1.1 (1770 K)	0.563 (954 K)	0.361	0.303	0.276	0.45	N/m	
Temperature coefficient			-0.068×10^3		−0.095		N/(m K)	
Coefficient of linear thermal expansion α	11.5×10^{-6}	26.1×10^{-6}	22×10^{-6}	23×10^{-6}	20.7×10^{-6}	20.2×10^{-6}	1/K	
Sound velocity, solid, transverse	8330	3170	2210	1520	1160		m/s	
Sound velocity, solid, longitudinal	12 720	5700	4180	2780	2080		m/s	
Compressibility κ	0.765×10^{-5}	2.88×10^{-5}	5.73×10^{-5} [a]	7.97×10^{-5}	10.0×10^{-5}		1/MPa	Volume compressibility
Elastic modulus E	286	44.4	19.6[a]	15.7	12.8	13.2[b]	GPa	
Shear modulus G	133	16.9	7.85[a]	6.03	4.84		GPa	
Poisson number μ	0.12	0.28	0.31[a]	0.28	0.28			
Elastic compliance s_{11}	3.45	22.0	104	218	157		1/TPa	
Elastic compliance s_{33}	2.87	19.7					1/TPa	
Elastic compliance s_{44}	6.16	60.9	71	135	105		1/TPa	
Elastic compliance s_{12}	−0.28	−7.8	−42	−90	−61		1/TPa	
Elastic compliance s_{13}	−0.05	−5.0					1/TPa	
Elastic stiffness c_{11}	292	59.3	22.8	10.94	12.6		GPa	
Elastic stiffness c_{33}	349	61.5					GPa	
Elastic stiffness c_{44}	163	16.4	14	7.41	9.5		GPa	
Elastic stiffness c_{12}	24	25.7	16.0	7.69	8.0		GPa	
Elastic stiffness c_{13}	6	21.4					GPa	
Tensile strength	228–352	90					GPa	
Vickers hardness	1670		130	140	42			
Brinell hardness	1060–1300	300–500						At 293 K

[a] At 348 K.

[b] Estimated.

Table 2.1–9B(c) Elements of Group IIA (CAS notation), or Group 2 (new IUPAC notation). Thermal and thermodynamic properties

Element name	Beryllium	Magnesium	Calcium	Strontium	Barium	Radium		
Chemical symbol	Be	Mg	Ca	Sr	Ba	Ra		
Atomic number Z	4	12	20	38	56	88		
							Units	Remarks
Thermal conductivity λ	200	171	200	35.3	18.4	18.6	W/(m K)	
Molar heat capacity c_p	16.44	24.895	25.940	26.4	28.09		J/(mol K)	At 298 K
Standard entropy S^0	9.500	32.671	41.588	55.694	62.500	69.000	J/(mol K)	At 298 K and 100 kPa
Enthalpy difference $H_{298} - H_0$	1.9500	4.9957	5.7360	6.5680	6.9100	7.2000	kJ/mol	
Melting temperature T_m	1560.0	923.00	1115.0	1050.0	1000.0	969.00	K	
Enthalpy change ΔH_m	7.8950	8.4768	8.5395	7.4310	7.1190	7.7000	kJ/mol	
Entropy change ΔS_m	5.061	9.184	7.659	7.077	7.119	7.946	J/(mol K)	
Relative volume change ΔV_m		0.042	0.047					$(V_l - V_s)/V_l$ at T_m
Boiling temperature T_b	2741	1366	1773	1685	2118		K	
Enthalpy change ΔH_b	291.58	127.4	153.6	137.19	141.5	136.7	kJ/mol	

Table 2.1–9B(d) Elements of Group IIA (CAS notation), or Group 2 (new IUPAC notation). Electronic, electromagnetic, and optical properties

Element name	Beryllium	Magnesium	Calcium	Strontium	Barium	Radium		
Chemical symbol	Be	Mg	Ca	Sr	Ba	Ra		
Atomic number Z	4	12	20	38	56	88		
							Units	Remarks
Characteristics	Light metal	Light metal	Metal	Soft metal	Soft metal	Metal		
Electrical resistivity ρ_s	28	39.4	31.6	303	500		nΩ m	At RT
Temperature coefficient	90.0×10^{-4}	41.2×10^{-4}	41.7×10^{-4}	38.2×10^{-4}	64.9×10^{-4}		1/K	
Pressure coefficient	-1.6×10^{-9}	-4.7×10^{-9}	15.2×10^{-9}	5.56×10^{-9}	-3.0×10^{-9}		1/hPa	
Electrical resistivity ρ_l	27.4						nΩ m	
Resistivity ratio		1.78						ρ_l/ρ_s at T_m
Hall coefficient R	2.4×10^{-10}	-0.83×10^{-10}					$m^3/(A s)$	At 300 K, $B = 1$ T
Thermoelectric coefficient		−0.4	−8.2				μV/K	
Electronic work function	4.98	3.97	2.87	2.74	2.56		V	
Thermal work function	3.37	3.46	2.76	2.35	2.29		V	
Molar magnetic susceptibility χ_{mol}, solid (SI)	-113×10^{-6}	165×10^{-6}	503×10^{-6}	1.16×10^{-3}	259×10^{-6}		cm^3/mol	At 295 K
Molar magnetic susceptibility χ_{mol}, solid (cgs)	-9.0×10^{-6}	13.1×10^{-6}	40.0×10^{-6}	92.0×10^{-6}	20.6×10^{-6}		cm^3/mol	At 295 K
Mass magnetic susceptibility χ_{mass}, solid (SI)	-13×10^{-6}	6.8×10^{-6}	14×10^{-6}	13.2×10^{-6}	1.9×10^{-6}		cm^3/g	At 295 K
Refractive index n, solid		0.37						$\lambda = 589$ nm

Table 2.1–9C Elements of Group IIA (CAS notation), or Group 2 (new IUPAC notation). Allotropic and high-pressure modifications

Element	Beryllium		Calcium		Strontium				
Modification	α-Be	β-Be	α-Ca	γ-Ca	α-Sr	β-Sr	γ-Sr	Sr-II	
									Units
Crystal system, Bravais lattice	hex, cp	cub, bc	cub, fc	cub, bc	cub, fc	hex, cp	cub, bc	cub, bc	
Structure type	Mg	W	Cu	W	Cu	Mg	W	W	
Lattice constant *a*	228.57	255.15	558.84	448.0	608.4	428	487	443.7	pm
Lattice constant *c*	358.39					705			pm
Space group	$P6_3/mmc$	$Im\bar{3}m$	$Fm\bar{3}m$	$Im\bar{3}m$	$Fm\bar{3}m$	$P6_3/mmc$	$Im\bar{3}m$	$Im\bar{3}m$	
Schoenflies symbol	D_{6h}^4	O_h^9	O_h^5	O_h^9	O_h^5	D_{6h}^4	O_h^9	O_h^9	
Strukturbericht type	A3	A2	A1	A2	A1	A3	A2	A2	
Pearson symbol	hP2	cI2	cF4	cI2	cF4	hP2	cI2	cI2	
Number *A* of atoms per cell	2	2	4	2	4	2	2	2	
Coordination number	12	8	12	8	12	12	8	8	
Shortest interatomic distance, solid	222	221	393	388	430		422		pm
Range of stability	RT	> 1523 K	RT	> 1010 K	RT	> 486 K	> 878 K	> 3.5 GPa	

Table 2.1–9D Elements of Group IIA (CAS notation), or Group 2 (new IUPAC notation). Ionic radii (determined from crystal structures)

Element	Beryllium	Magnesium	Calcium	Strontium	Barium	Radium	
Ion	Be^{2+}	Mg^{2+}	Ca^{2+}	Sr^{2+}	Ba^{2+}	Ra^{2+}	
Coordination number							Units
4	27	57					pm
6	45	72	100	118	135		pm
8		89	112	126	142	148	pm
10			123	136			pm
12			134	144	161	170	pm

Table 2.1-10A Elements of Group IIB (CAS notation), or Group 12 (new IUPAC notation). Atomic, ionic, and molecular properties (see Table 2.1-10D for ionic radii)

Element name	Zinc	Cadmium	Mercury		
Chemical symbol	Zn	Cd	Hg		
Atomic number Z	30	48	80		
				Units	Remarks
Relative atomic mass A (atomic weight)	65.39(2)	112.411(8)	200.59(2)		
Abundance in lithosphere	80×10^{-6}	0.18×10^{-6}	0.5×10^{-6}		Mass ratio
Abundance in sea	4.9×10^{-9}	1×10^{-10}	3×10^{-11}		Mass ratio
Atomic radius r_{cov}	125	141	144	pm	Covalent radius
Atomic radius r_{met}	134	149	162	pm	Metallic radius, CN = 12
Atomic radius r_{vdW}	140	160	150	pm	van der Waals radius
Electron shells	–LMN	–MNO	–NOP		
Electronic ground state	$^1S_0$	$^1S_0$	$^1S_0$		
Electronic configuration	$[Ar]3d^{10}4s^2$	$[Kr]4d^{10}5s^2$	$[Xe]4f^{14}5d^{10}6s^2$		
Oxidation states	2+	2+	2+, 1+		
Electron affinity	Not stable	Not stable	Not stable	eV	
Electronegativity χ_A	1.66	1.46	(1.44)		Allred and Rochow
1st ionization energy	9.39405	8.99367	10.43750	eV	
2nd ionization energy	17.96440	16.90832	18.756	eV	
3rd ionization energ	39.723	37.48	34.2	eV	
4th ionization energy	59.4			eV	
Standard electrode			+0.7973	V	Reaction type $Hg^{2+}+2e^-=2Hg$
potential E^0	−0.7618	−0.403		V	Reaction type $Cu^{2+}+2e^-=Cu$

Table 2.1-10B(a) Elements of Group IIB (CAS notation), or Group 12 (new IUPAC notation). Crystallographic properties (see Table 2.1-10 for allotropic and high-pressure modifications)

Element name	Zinc	Cadmium	Mercury		
Chemical symbol	Zn	Cd	Hg		
Atomic number Z	30	48	80		
				Units	Remarks
Modification			α-Hg		
Crystal system, Bravais lattice	hex	hex	trig, R		
Structure type	Mg	Mg	α-Hg		
Lattice constant a	266.44	297.88	300.5 (225 K)	pm	At 298 K
Lattice constant c	494.94	561.67		pm	At 298 K
Lattice angle α			70.53 (225 K)	deg	
Space group	$P6_3/mmc$	$P6_3/mmc$	$R\bar{3}m$		
Schoenflies symbol	D_{6h}^4	D_{6h}^4	D_{3d}^5		
Strukturbericht type	A3	A3	A 10		
Pearson symbol	hP2	hP2	hR1		
Number A of atoms per cell	2	2	1		
Coordination number	6+6	6+6	6+6		
Shortest interatomic distance, solid	266	297	299 (at 234 K)	pm	At 293 K

Table 2.1-10B(b) Elements of Group IIB (CAS notation), or Group 12 (new IUPAC notation). Mechanical properties

Element name	Zinc	Cadmium	Mercury		
Chemical symbol	Zn	Cd	Hg		
Atomic number Z	30	48	80		
				Units	Remarks
Density ϱ, solid	7.14	8.65		g/cm^3	
Density ϱ, liquid	6.570	8.02	13.53 (25 °C)	g/cm^3	
Molar volume V_{mol}	9.17	13.00	14.81	cm^3/mol	
Viscosity η, liquid	2.95	2.29	1.55	mPa s	At T_m
Surface tension, liquid	0.816	0.564	0.476	N/m	
Temperature coefficient	0.25×10^{-3}	0.39×10^{-3}	-0.20×10^{-3}	N/(m K)	
Coefficient of linear thermal expansion α	25.0×10^{-6}	29.8×10^{-6}	18.1	1/K	At 298.15 K
Sound velocity, liquid	2790	2200[a]	1451	m/s	12 MHz[a]
Sound velocity, solid, transverse	2290	1690		m/s	
Sound velocity, solid, longitudinal	3890	2980		m/s	
Compressibility κ	1.65×10^{-5}	2.14×10^{-5}	3.77×10^{-5}	1/MPa	Volume compressibility
Elastic modulus E	92.7	62.3	25	GPa	
Shear modulus G	34.3	24.5		GPa	
Poisson number μ	0.29	0.30			
Elastic compliance s_{11}	8.22	12.4	154 (83 K)	1/TPa	
Elastic compliance s_{33}	27.7	34.6	45 (83 K)	1/TPa	
Elastic compliance s_{44}	25.3	53.1	151 (83 K)	1/TPa	
Elastic compliance s_{12}	0.60	−1.2	−119 (83 K)	1/TPa	
Elastic compliance s_{13}	−7.0	−9.1	−21 (83 K)	1/TPa	
Elastic compliance s_{14}			−100 (83 K)	1/TPa	
Elastic stiffness c_{11}	165	114.1	36.0 (83 K)	GPa	
Elastic stiffness c_{33}	61.8	49.9	50.5 (83 K)	GPa	
Elastic stiffness c_{44}	39.6	19.0	12.9 (83 K)	GPa	
Elastic stiffness c_{12}	31.1	41.0	28.9 (83 K)	GPa	
Elastic stiffness c_{13}	50.0	40.3	30.3 (83 K)	GPa	
Elastic stiffness c_{14}			4.7 (83 K)	GPa	
Tensile strength	20–40	71		MPa	293 K
Brinell hardness	280–330	180–230			

[a] At 594 K

Table 2.1–10B(c) Elements of Group IIB (CAS notation), or Group 12 (new IUPAC notation). Thermal and thermodynamic properties

Element name	Zinc	Cadmium	Mercury		
Chemical symbol	Zn	Cd	Hg		
Atomic number Z	30	48	80		
				Units	Remarks
State	Solid	Solid	Liquid		
Thermal conductivity λ	121	96.8	8.34	W/(m K)	At 300 K
Molar heat capacity c_p	25.44	25.98	27.983	J/(mol K)	At 298 K
Standard entropy S^0	41.631	51.800	75.900	J/(mol K)	At 298 K and 100 kPa
Enthalpy difference $H_{298} - H_0$	5.6570	6.2470	9.3420	kJ/mol	
Melting temperature T_m	692.68	594.22	234.32	K	
Enthalpy change ΔH_m	7.3220	6.1923	2.2953	kJ/mol	
Entropy change ΔS_m	10.571	10.421	9.796	J/(mol K)	
Relative volume change	0.0730	0.0474	0.037		$(V_l - V_s)/V_l$ at T_m
Boiling temperature T_b	1180	1040	629	K	
Enthalpy change ΔH_b	115.3	97.40	59.2	kJ/mol	
Critical temperature T_c			1750	K	
Critical pressure p_c			167	MPa	
Critical density ϱ_c			5.700	g/cm^3	

Table 2.1-10B(d) Elements of Group IIB (CAS notation), or Group 12 (new IUPAC notation). Electronic, electromagnetic, and optical properties

Element name	Zinc	Cadmium	Mercury		
Chemical symbol	**Zn**	**Cd**	**Hg**		
Atomic number *Z*	**30**	**48**	**80**		
				Units	**Remarks**
Characteristics	Ductile metal	Soft metal	Noble metal		
Electrical resistivity ρ_s	54.3	68		nΩ m	At 293 K
Temperature coefficient	41.7×10^{-4}	46.2×10^{-4}		1/K	
Pressure coefficient	-6.3×10^{-9}	-7.32×10^{-9}		1/hPa	
Electrical resistivity ρ_l	326	337	958 (298 K)	nΩ m	
Resistivity ratio	2.1	1.97	3.74–4.94		ρ_l/ρ_s at T_m
Superconducting critical temperature T_{crit}	0.88	0.55	4.15 K	K	α-Hg
Superconducting critical field H_{crit}	53	30	412	Oe	α-Hg
Hall coefficient *R*	0.63×10^{-10}	0.589×10^{-10}	-0.73×10^{-10}	$m^3/(A\,s)$	$B = 0.3$–2.2 T 298 K
Thermoelectric coefficient	2.9	2.8	−3.4	μV/K	
Electronic work function	4.22	4.04		V	
Thermal work function	3.74	3.92		V	
Molar magnetic susceptibility χ_{mol}, solid (SI)	-115×10^{-6}	-248×10^{-6}	-303×10^{-6} (234 K)	cm^3/mol	At 295 K
Molar magnetic susceptibility χ_{mol}, solid (cgs)	-9.15×10^{-6}	-19.7×10^{-6}	-24.1×10^{-6} (234 K)	cm^3/mol	At 295 K
Molar magnetic susceptibility χ_{mol}, liquid (SI)			-421×10^{-6}	cm^3/mol	At 295 K
Molar magnetic susceptibility χ_{mol}, liquid (cgs)			-33.5×10^{-6}	cm^3/mol	At 295 K
Mass magnetic suscepti-, bility χ_{mass}, solid (SI)	-1.38×10^{-6}	-2.21×10^{-6}		cm^3/g	At 295 K
Mass magnetic susceptibility χ_{mass}, liquid (SI)	-1.32×10^{-6}	-1.83×10^{-6}		cm^3/g	
Refractive index $(n-1)$, gas			$+1882 \times 10^{-6}$ (Hg_2 vapor)		
Refractive index *n*, solid	1.19 (λ = 550 nm)	1.8 (578 nm)			

Table 2.1-10C Elements of Group IIB (CAS notation), or Group 12 (new IUPAC notation). Allotropic and high-pressure modifications

Element	Mercury		
Modification	α-Hg	β-Hg	
			Units
Crystal system, Bravais lattice	trig, R	tetr	
Structure type	α-Hg	In	
Lattice constant a	300.5 (225 K)	399.5	pm
Lattice constant c		282.5	pm
Lattice angle α	70.53 (225 K)		
Space group	$R\bar{3}m$	$I4/mmm$	
Schoenflies symbol	D_{3d}^5	D_{4h}^{17}	
Strukturbericht type	A10		
Pearson symbol	hR1	tI2	
Number A of atoms per cell	1	2	
Coordination number	6+6	2+8	
Shortest interatomic distance, solid	299	283	pm
Range of stability	< 234.2 K	77 K, high pressure	

Table 2.1-10D Elements of Group IIB (CAS notation), or Group 12 (new IUPAC notation). Ionic radii (determined from crystal structures)

Element	Zinc	Cadmium	Mercury		
Ion	Zn^{2+}	Cd^{2+}	Hg^{+}	Hg^{2+}	
Coordination number					Units
2				69	pm
4	60	78		96	pm
6	74	95	119	102	pm
8	90	110		114	pm
12		131			pm

Table 2.1-11A Elements of Group IIIA (CAS notation), or Group 13 (new IUPAC notation). Atomic, ionic, and molecular properties (see Table 2.1-11D for ionic radii)

Element name	Boron	Aluminium	Gallium	Indium	Thallium		
Chemical symbol	**B**	**Al**	**Ga**	**In**	**Tl**		
Atomic number Z	**5**	**13**	**31**	**49**	**81**		
						Units	**Remarks**
Relative atomic mass A (atomic weight)	10.811(7)	26.981538(2)	69.723(1)	114.818(3)	204.3833(2)		
Abundance in lithosphere	10×10^{-6}	$81\,300 \times 10^{-6}$	15×10^{-6}	0.06×10^{-6}	0.3×10^{-6}		Mass ratio
Abundance in sea	4.4×10^{-6}	0.002×10^{-6}	3×10^{-11}	1×10^{-13}	1×10^{-11}		Mass ratio
Atomic radius r_{cov}	81	125	125	150	155	pm	Covalent radius
Atomic radius r_{met}	89	143	153	167	170	pm	Metallic radius, CN = 12
Atomic radius r_{vdW}		205	190	190	200	pm	van der Waals radius
Electron shells	KL	KLM	–LMN	–MNO	–NOP		
Electronic ground state	$^2P_{1/2}$	$^2P_{1/2}$	$^2P_{1/2}$	$^2P_{1/2}$	$^2P_{1/2}$		
Electronic configuration	$[He]2s^22p^1$	$[Na]3s^23p^1$	$[Ar]3d^{10}4s^24p^1$	$[Kr]4d^{10}5s^25p^1$	$[Xe]4f^{14}5d^{10}6s^26p^1$		
Oxidation states	3+	3+	3+	3+	3+, 1+		
Electron affinity	0.277	0.441	0.3	0.3	0.2	eV	
Electronegativity χ_A	2.01	1.47	1.82	1.49	1.44		Allred and Rochow
1st ionization energy	8.29803	5.98577	5.99930	5.78636	6.10829	eV	
2nd ionization energy	25.15484	18.82856	20.5142	18.8698	20.428	eV	
3rd ionization energy	37.93064	28.44765	30.71	28.03	29.83	eV	
4th ionization energy	259.37521	119.992	64	54		eV	
Standard electrode potential E^0					−0.336	V	Reaction type $Tl^+ + e^- = Tl$
		−1.662	−0.560	−0.338		V	Reaction type $Al^{3+} + 3e^- = Al$
					+1.252	V	Reaction type $Tl^{3+} + 2e^- = Tl^+$

Table 2.1-11B(a) Elements of Group IIIA (CAS notation), or Group 13 (new IUPAC notation). Crystallographic properties (see Table 2.1-11C for allotropic and high-pressure modifications)

Element name	**Boron**	**Aluminium**	**Gallium**	**Indium**	**Thallium**		
Chemical symbol	**B**	**Al**	**Ga**	**In**	**Tl**		
Atomic number Z	**5**	**13**	**31**	**49**	**81**		
						Units	**Remarks**
Modification	β-B	α-Al	α-Ga				
Crystal system, Bravais lattice	trig, R	cub, fc	orth, C	tetr, I	hex		
Structure type	β-B	Cu	α-Ga	In	Mg		
Lattice constant a		361.49	451.92	459.90	345.63	pm	
Lattice constant b			765.86			pm	
Lattice constant c			452.58	494.70	552.63	pm	
Space group	$R\bar{3}m$	$Fm\bar{3}m$	$Cmca$	$I4/mmm$	$P6_3/mmc$		
Schoenflies symbol	D_{3d}^5	O_h^5	D_{2h}^{18}	D_{4h}^{17}	D_{6h}^4		
Strukturbericht type		A1	A11	A6	A3		
Pearson symbol	hR105	cF4	oC8	tI2	hP2		
Number A of atoms per cell	105	4	8	2	2		
Coordination number		12	1+2+2+2	4+8			
Shortest interatomic distance, solid	162–192	286	247	325		pm	

Table 2.1–11B(b) Elements of Group IIIA (CAS notation), or Group 13 (new IUPAC notation). Mechanical properties

Element name	Boron	Aluminium	Gallium	Indium	Thallium		
Chemical symbol	B	Al	Ga	In	Tl		
Atomic number Z	5	13	31	49	81		
						Units	Remarks
Density ϱ, solid	2.46	2.70	5.91	7.31	11.85	g/cm^3	
Density ϱ, liquid		2.39	6.200	6.990	11.29	g/cm^3	At T_m
Molar volume V_{mol}	4.62	10.00	11.81	15.71	17.24	cm^3/mol	
Viscosity η, liquid		1.38	1.70	1.65		mPa s	At T_m
Surface tension γ, liquid		0.860	0.718	0.556	0.447	N/m	At T_m
Temperature coefficient		-0.135×10^{-3}		-0.09×10^{-3}	-0.07×10^{-3}	N/(m K)	
Coefficient of linear thermal expansion α	5×10^{-6}	23.03×10^{-6}	18.3×10^{-6}	33×10^{-6}	28×10^{-6}	1/K	
Sound velocity, liquid		4650	2740	2215		m/s	At 12 MHz
Sound velocity, solid, transverse		3130	750	710	480	m/s	
Sound velocity, solid, longitudinal		6360	3030	2460	1630	m/s	
Compressibility κ	0.539×10^{-5}	1.33×10^{-5}	1.96×10^{-5}	2.70×10^{-5}	3.41×10^{-5}	1/MPa	Volume compressibility
Elastic modulus E	178[a]	70.2	9.81	10.6	7.89	GPa	
Shear modulus G		27.8	6.67	3.68	2.67	GPa	
Poisson number μ		0.34	0.47	0.45	0.45		
Elastic compliance s_{11}		16.0	12.2	148.8	104	1/TPa	
Elastic compliance s_{22}			14.0			1/TPa	
Elastic compliance s_{33}			8.49	196.2	31.1	1/TPa	
Elastic compliance s_{44}		35.3	28.6	153.7	139	1/TPa	
Elastic compliance s_{55}			23.9			1/TPa	
Elastic compliance s_{66}			24.8	83.2		1/TPa	
Elastic compliance s_{12}		−5.8	−4.4	−46.0	−83	1/TPa	
Elastic compliance s_{13}			−1.7	−94.5	−11.6	1/TPa	
Elastic compliance s_{23}			−2.4			1/TPa	

[a] Estimated.

Table 2.1–11B(b) Elements of Group IIIA (CAS notation), or Group 13 (new IUPAC notation). Mechanical properties , cont.

Element name	Boron	Aluminium	Gallium	Indium	Thallium		
Chemical symbol	B	Al	Ga	In	Tl		
Atomic number Z	5	13	31	49	81		
						Units	Remarks
Elastic stiffness c_{11}	467	108	100	45.1	41.9	GPa	
Elastic stiffness c_{22}			90.2			GPa	
Elastic stiffness c_{33}	473		135	44.6	54.9	GPa	
Elastic stiffness c_{44}	198	28.3	35.0	6.51	7.20	GPa	
Elastic stiffness c_{55}			41.8			GPa	
Elastic stiffness c_{66}			40.3	12.0		GPa	
Elastic stiffness c_{12}	241	62	37	40.0	36.6	GPa	
Elastic stiffness c_{13}			33	41.0	29.9	GPa	
Elastic stiffness c_{23}			31			GPa	
Elastic stiffness c_{14}	15.1					GPa	
Tensile strength	16–24[a]	90–100			8.9	MPa	
Vickers hardness	49 000	167			9		At 293 K
Brinell hardness				0.9			
Mohs hardness	9.3		1.5–2.5				

[a] Amorphous.

Table 2.1–11B(c) Elements of Group IIIA (CAS notation), or Group 13 (new IUPAC notation). Thermal and thermodynamic properties

Element name	Boron	Aluminium	Gallium	Indium	Thallium		
Chemical symbol	B	Al	Ga	In	Tl		
Atomic number Z	5	13	31	49	81		
						Units	Remarks
State	Crystalline						
Thermal conductivity λ	27.0	237	33.5	81.6	46.1	W/(m K)	
Thermal conductivity (liquid) λ_l		90				W/(m K)	At T_m
Molar heat capacity c_p	11.20	24.392	26.15	26.732	26.32	J/(mol K)	At 298 K
Standard entropy S^0	5.900	28.300	40.727	57.650	64.300	J/(mol K)	At 298 K and 100 kPa
Enthalpy difference $H_{298} - H_0$	1.2220	4.5400	5.5720	6.6100	6.832	kJ/mol	
Melting temperature T_m	2348.00	933.47	302.91	429.75	577.00	K	
Enthalpy change ΔH_m	50.200	10.7110	5.5898	3.2830	4.1422	kJ/mol	
Entropy change ΔS_m	21.380	11.474	18.454	7.639	7.179	J/(mol K)	
Volume change ΔV_m		0.065	−0.034	0.025	0.0323		$(V_l - V_s)/V_l$ at T_m
Boiling temperature T_b	4138	2790	2478	2346	1746	K	
Enthalpy change ΔH_b	480.5	294.0	258.7	231.45	164.1	kJ/mol	

Table 2.1–11B(d) Elements of Group IIIA (CAS notation), or Group 13 (new IUPAC notation). Electronic, electromagnetic, and optical properties

Element name	Boron	Aluminium	Gallium	Indium	Thallium		
Chemical symbol	**B**	**Al**	**Ga**	**In**	**Tl**		
Atomic number Z	**5**	**13**	**31**	**49**	**81**		
						Units	**Remarks**
Characteristics	Semiconductor	Light metal	Soft metal	Soft metal	Soft metal, toxic		
Electrical resistivity ρ_s	6500×10^9 [a]	25.0	136	80.0	150	nΩ m	[a] At 300 K
Temperature coefficient		46×10^{-4}	39.6×10^{-4}	49.0×10^{-4}	51.7×10^{-4}	1/K	
Pressure coefficient		-4.06×10^{-9}	-2.47×10^{-9}	-12.2×10^{-9}	-3.4×10^{-9}	1/hPa	
Electrical resistivity ρ_l		200	258	331	740	nΩ m	
Resistivity ratio at T_m		1.64	1.9	2.18			ρ_l/ρ_s at T_m
Superconducting critical temperature T_{cr}		1.2	1.09	3.4	2.4	K	
Superconducting critical field H_{cr}		99	51	293	171	Oe	
Electronic band gap ΔE	1.5					eV	
Electronic work function	4.79	4.28	4.35	4.08	4.05	V	
Thermal work function	5.71	3.74	4.12	4.0	3.76	V	
Intrinsic charge carrier concentration	$+5\times10^{14}$					$1/cm^3$	At 430 K
Electron mobility	1					$cm^2/(V s)$	
Hole mobility	55					$cm^2/(V s)$	
Hall coefficient R		-0.343×10^{-10}	-0.63×10^{-10}	-0.24×10^{-10}	0.240×10^{-10}	$m^3/(A s)$	B=1.0–1.8 T 300 K
Thermoelectric coefficient		−0.6		2.4	0.4	μV/K	
Dielectric constant ε, solid	13 (0.5 MHz)						
Molar magnetic susceptibility χ_{mol}, solid (SI)	-84.2×10^{-6}	207×10^{-6}	-271×10^{-6}	-128×10^{-6}	-628×10^{-6}	cm^3/mol	At 295 K
Molar magnetic susceptibility χ_{mol}, solid (cgs)	-6.7×10^{-6}	16.5×10^{-6}	-21.6×10^{-6}	-10.2×10^{-6}	-50.0×10^{-6}	cm^3/mol	At 295 K
Mass magnetic susceptibility χ_{mass}, solid (SI)		$+7.9\times10^{-6}$	-3.9×10^{-6}	-1.4×10^{-6}	-3.13×10^{-6}	cm^3/g	At 295 K
Mass magnetic susceptibility χ_{mass}, liquid (SI)					-1.72×10^{-6}	cm^3/g	At T_m
Refractive index n, solid	3.2 (λ=1 μm)						

[a] At 300 K.

Table 2.1-11C Elements of Group IIIA (CAS notation), or Group 13 (new IUPAC notation). Allotropic and high-pressure modifications

Element	Aluminium		Gallium			Thallium			
Modification	α-Al	β-Al	α-Ga	β-Ga	γ-Ga	α-Tl	β-Tl	γ-Tl	
									Units
Crystal system, Bravais lattice	cub, fc	hex, cp	orth, C	tetr	orth	hex, cp	cub, bc	cub, fc	
Structure type	Cu	Mg	α-Ga	In	γ-Ga	Mg	W	Cu	
Lattice constant a	404.96	269.3	451.92	280.8	1059.3	345.63	387.9		pm
Lattice constant b			765.86		1352.3				pm
Lattice constant c		439.8	452.58	445.8	520.3	552.63			pm
Space group	$Fm\bar{3}m$	$P6_3/mmc$	$Cmca$	$I4/mmm$	$Cmcm$	$P6_3/mmc$	$Im\bar{3}m$	$Fm\bar{3}m$	
Schoenflies symbol	O_h^5	D_{6h}^4	D_{2h}^{18}		D_{2h}^{17}	D_{6h}^4	O_h^9	O_h^5	
Strukturbericht type	A1	A3	A11	A6		A3	A2	A1	
Pearson symbol	cF4	hP2	oC8	tI2	oC40	hP2	cI2	cF4	
Number A of atoms per cell	4	2	8	2	40	2	2	4	
Coordination number	12	12	1+2+2+2	4+8			8	12	
Shortest interatomic distance, solid	286		247	281	260–308		336		
Range of stability	RTP	> 20.5 GPa	RTP	> 1.2 GPa	220 K, > 3.0 GPa	RTP	> 503 K	High pressure	

Table 2.1-11D Elements of Group IIIA (CAS notation), or Group 13 (new IUPAC notation). Ionic radii (determined from crystal structures)

Element	Aluminium	Gallium	Indium	Thallium		
Ion	Al^{3+}	Ga^{3+}	In^{3+}	Tl^{+}	Tl^{3+}	
Coordination number						**Units**
4	39	47	62		75	pm
5	48					pm
6	54	62	80	150	89	pm
8				159	98	pm
12				170		pm

Table 2.1–12A Elements of Group IIIB (CAS notation), or Group 3 (new IUPAC notation). Atomic, ionic, and molecular properties (see Table 2.1-12D for ionic radii)

Element name	**Scandium**	**Yttrium**	**Lanthanum**	**Actinium**		
Chemical symbol	**Sc**	**Y**	**La**	**Ac**		
Atomic number Z	**21**	**39**	**57**	**89**		
					Units	**Remarks**
Characteristics				Radioactive		
Relative atomic mass A (atomic weight)	44.955910(8)	88.90585(2)	138.9055(2)	[227]		
Abundance in lithosphere	5×10^{-6}	28.1×10^{-6}	18.3×10^{-6}			Mass ratio
Abundance in sea	6×10^{-13}	3×10^{-12}	3×10^{-12}			Mass ratio
Atomic radius r_{cov}	144	162	169		pm	Covalent radius
Atomic radius r_{met}	166	178	187	188	pm	Metallic radius, CN = 12
Electron shells	–LMN	–MNO	–NOP	–OPQ		
Electronic ground state	$^2D_{3/2}$	$^2D_{3/2}$	$^2D_{3/2}$	$^2D_{3/2}$		
Electronic configuration	$[Ar]3d^14s^2$	$[Kr]4d^15s^2$	$[Xe]5d^16s^2$	$[Rn]6d^17s^2$		
Oxidation states	3+	3+	3+	3+		
Electron affinity	0.188	0.307	0.5		eV	
Electronegativity χ_A	1.20	1.11	1.08	(1.00)		Allred and Rochow
1st ionization energy	6.56144	6.217	5.5770	5.17	eV	
2nd ionization energy	12.79967	12.24	11.060	12.1	eV	
3rd ionization energy	24.75666	20.52	19.1773		eV	
4th ionization energy	73.4894	60.597	49.95		eV	
Standard electrode potential E^0			−2.522		V	Reaction type $La^{3+} + 3e^- = La$

Table 2.1-12B(a) Elements of Group IIIB (CAS notation), or Group 3 (new IUPAC notation). Crystallographic properties (see Table 2.1-12C for allotropic and high-pressure modifications)

Element name	**Scandium**	**Yttrium**	**Lanthanum**	**Actinium**		
Chemical symbol	**Sc**	**Y**	**La**	**Ac**		
Atomic number Z	**21**	**39**	**57**	**89**		
					Units	**Remarks**
Modification	α-Sc	α-Y	α-La			
Crystal system, Bravais lattice	hex	hex	hex			
Structure type	Mg	Mg	α-La			
Lattice constant a	330.88	364.82	377.40		pm	
Lattice constant c	526.80	573.18	1217.1		pm	
Space group	$P6_3/mmc$	$P6_3/mmc$	$P6_3/mmc$			
Schoenflies symbol	D_{6h}^4	D_{6h}^4	D_{6h}^4			
Strukturbericht type	A3	A3	A3′			
Pearson symbol	hP2	hP2	hP4			
Number A of atoms per cell	2	2	4			
Coordination number	12	6+6	12			
Shortest interatomic distance, solid	166		364		pm	

Table 2.1-12B(b) Elements of Group IIIB (CAS notation), or Group 3 (new IUPAC notation). Mechanical properties

Element name	**Scandium**	**Yttrium**	**Lanthanum**	**Actinium**		
Chemical symbol	**Sc**	**Y**	**La**	**Ac**		
Atomic number Z	**21**	**39**	**57**	**89**		
					Units	**Remarks**
Density ϱ, solid	2.989	4.50	6.70	10.07	g/cm^3	
Molar volume V_{mol}	15.04	19.89	22.60		cm^3/mol	
Surface tension, liquid	0.9	0.9	0.71		N/m	
Coefficient of linear thermal expansion α	10.0×10^{-6}	10.6×10^{-6}	4.9×10^{-6}		1/K	
Sound velocity, solid, transverse		2420	1540		m/s	
Sound velocity, solid, longitudinal		4280	2770		m/s	
Compressibility κ	2.22×10^{-5}	2.62×10^{-5}	3.96×10^{-5}		1/MPa	Volume compressibility
Elastic modulus E	75.2	66.3	39.2	25[a]	GPa	
Shear modulus G	29.7	25.5	14.9		GPa	
Poisson number μ	0.28	0.27	0.28			
Elastic compliance s_{11}	12.5	15.4	51.7 [b]		1/TPa	
Elastic compliance s_{33}	10.6	14.4			1/TPa	
Elastic compliance s_{44}	36.1	41.1	55.7 [b]		1/TPa	
Elastic compliance s_{12}	−4.3	−5.1	−19.2 [b]		1/TPa	
Elastic compliance s_{13}	−2.2	−2.7			1/TPa	
Elastic stiffness c_{11}	99.3	77.9	34.5 [b]		GPa	
Elastic stiffness c_{33}	107	76.9			GPa	
Elastic stiffness c_{44}	27.7	24.3	18.0 [b]		GPa	
Elastic stiffness c_{12}	39.7	29.2	20.4 [b]		GPa	
Elastic stiffness c_{13}	29.4	20			GPa	
Tensile strength	256	250–380			MPa	
Vickers hardness	350	40	491			

[a] Estimated.

[b] For lanthanum in its metastable fcc phase at room temperature.

Table 2.1-12B(c) Elements of Group IIIB (CAS notation), or Group 3 (new IUPAC notation). Thermal and thermodynamic properties

Element name	**Scandium**	**Yttrium**	**Lanthanum**	**Actinium**		
Chemical symbol	**Sc**	**Y**	**La**	**Ac**		
Atomic number Z	**21**	**39**	**57**	**89**		
					Units	**Remarks**
Thermal conductivity λ	15.8	17.2	13.5		W/(m K)	
Molar heat capacity c_p	25.52	26.53	27.11	27.2	J/(mol K)	At 298 K
Standard entropy S^0	34.644	44.788	56.902	62.000	J/(mol K)	At 298 K and 100 kPa
Enthalpy difference $H_{298} - H_0$	5.2174	5.9835	6.6651	6.7000	kJ/mol	
Melting temperature T_m	1814.00	1795.15	1193.00	1323.00	K	
Enthalpy change ΔH_m	14.0959	11.3942	6.1965	12.000	kJ/mol	
Entropy change ΔS_m	7.771	6.347	5.194	9.070	J/(mol K)	
Relative volume change ΔV_m			0.006			$(V_l - V_s)/V_l$ at T_m
Boiling temperature T_b	3104	3611	3730	3473	K	
Enthalpy change ΔH_b	314.2	363.3	413.7	293	kJ/mol	

Table 2.1–12B(d) Elements of Group IIIB (CAS notation), or Group 3 (new IUPAC notation). Electronic, electromagnetic, and optical properties

Element name	Scandium	Yttrium	Lanthanum	Actinium		
Chemical symbol	Sc	Y	La	Ac		
Atomic number Z	21	39	57	89		
					Units	Remarks
Characteristics	Soft metal	Reactive metal	Very reactive metal			
Electrical resistivity ρ_s	505	550	540		nΩ m	At RT
Temperature coefficient	28.2×10^{-4}	27.1×10^{-4}	21.8×10^{-4}		1/K	
Pressure coefficient			-1.7×10^{-9}		1/hPa	
Electrical resistivity ρ_l			1350		nΩ m	
Superconducting critical temperature T_{crit}			5.0		K	
Superconducting critical field H_{crit}						
Hall coefficient R	-0.67×10^{-10}	-0.770×10^{-10}	-0.8×10^{-10}		$m^3/(A\,s)$	At 293 K, $B=0.5-1.0$ T
Thermoelectric coefficient	−3.6	2.2			μV/K	
Electronic work function	3.5	3.1	3.5		V	
Thermal work function	3.23	3.07	3.3		V	
Molar magnetic susceptibility χ_{mol}, solid (SI)	3710×10^{-6}	2359×10^{-6}	$+1205\times10^{-6}$		cm^3/mol	At 295 K
Molar magnetic susceptibility χ_{mol}, solid (cgs)	295×10^{-6}	188×10^{-6}	$+95.9\times10^{-6}$		cm^3/mol	At 295 K
Mass magnetic susceptibility χ_{mass}, solid (SI)	88×10^{-6}	27.0×10^{-6}	$+11\times10^{-6}$		cm^3/g	

Table 2.1–12C Elements of Group IIIB (CAS notation), or Group 3 (new IUPAC notation). Allotropic and high-pressure modifications

Element	Scandium		Yttrium		Lanthanum				
Modification	α-Sc	β-Sc	α-Y	β-Y	α-La	β-La	γ-La	β′-La	
									Units
Crystal system, Bravais lattice	hex, cp	cub, bc	hex, cp	cub, bc	hex	cub, fc	cub, bc	cub, fc	
Structure type	Mg	W	Mg	W	α-La	Cu	W	Cu	
Lattice constant a	330.88		364.82		377.40	530.45	426.5	517	pm
Lattice constant c	526.80		573.18		1217.1				pm
Space group	$P6_3/mmc$	$Im\bar{3}m$	$P6_3/mmc$	$Im\bar{3}m$	$P6_3/mmc$	$Fm\bar{3}m$	$Im\bar{3}m$	$Fm\bar{3}m$	
Schoenflies symbol	D_{6h}^4	O_h^9	D_{6h}^4	O_h^9	D_{6h}^4	O_h^5	O_h^9	O_h^5	
Strukturbericht type	A3	A2	A3	A2	A3′	A1	A2	A1	
Pearson symbol	hP2	cI2	hP2	cI2	hP4	cF4	cI2	cF4	
Number A of atoms per cell	2	2	2	2	4	4	2	4	
Coordination number	12	8	6+6	8	12	12	8	12	
Shortest interatomic distance, solid				356	374	375	369		pm
Range of stability	RT	> 1607 K	RT	> 1752 K	RTP	> 613 K	> 1141 K	> 2.0 GPa	

Table 2.1–12D Elements of Group IIIB (CAS notation), or Group 3 (new IUPAC notation). Ionic radii (determined from crystal structures)

Element	Scandium	Yttrium	Lanthanum	Actinium	
Ion	Sc^{3+}	Y^{3+}	La^{3+}	Ac^{3+}	
Coordination number					Units
6	75	90	103	112	pm
8	87	102	116		pm
9		108			pm
10			127		pm
12			136		pm

Table 2.1–13A Elements of Group IVA (CAS notation), or Group 14 (new IUPAC notation). Atomic, ionic, and molecular properties (see Table 2.1-13D for ionic radii)

Element name	Carbon	Silicon	Germanium	Tin	Lead		
Chemical symbol Atomic number Z	C 6	Si 14	Ge 32	Sn 50	Pb 82		
						Units	Remarks
Relative atomic mass A (atomic weight)	12.0107(8)	28.0855(3)	72.61(2)	118.710(7)	207.2(1)		
Abundance in lithosphere	320×10^{-6}	$277\,200\times10^{-6}$	7×10^{-6}	40×10^{-6}	16×10^{-6}		Mass ratio
Abundance in sea	28×10^{-6}	2×10^{-6}	5×10^{-11}	1×10^{-11}	3×10^{-11}		Mass ratio
Atomic radius r_{cov}	77	117	122	140	154	pm	Covalent radius
Atomic radius r_{met}	91	132	137	220	175	pm	Metallic radius, CN = 12
Atomic radius r_{vdW}	170	210	234	158	200	pm	van der Waals radius
Electron shells	KL	KLM	–LMN	–MNO	–NOP		
Electronic ground state	$^3P_0$	$^3P_0$	$^3P_0$	$^3P_0$	$^3P_0$		
Electronic configuration	$1s^22s^22p^2$	$[Ne]3s^23p^2$	$[Ar]3d^{10}4s^24p^2$	$[Kr]4d^{10}5s^25p^2$	$[Xe]4f^{14}5d^{10}6s^26p^2$		
Oxidation states	4+, 4–, 2+	4+	4+	4+, 2+	4+, 2+		
Electron affinity	1.26	1.39	1.23	1.11	0.364	eV	
Electronegativity χ_A	2.50	1.74	2.02	1.72	1.55		Allred and Rochow
1st ionization energy	11.26030	8.15169	7.900	7.34381	7.41666	eV	
2nd ionization energy	24.38332	16.34585	15.93462	14.63225	15.0322	eV	
3rd ionization energy	47.8878	33.49302	34.2241	30.50260	31.9373	eV	
4th ionization energy	64.4939	45.14181	45.7131	40.73502	42.32	eV	
5th ionization energy	392.087	166.767	93.5	72.28	68.8	eV	
Standard electrode potential E^0				−0.137	−0.126	V	Reaction type $Sn^{2+}+2e^-=Sn$
				+0.151		V	Reaction type $Sn^{4+}+2e^-=Sn^{2+}$

Table 2.1–13B(a) Elements of Group IVA (CAS notation), or Group 14 (new IUPAC notation). Crystallographic properties (see Table 2.1-13C for allotropic and high-pressure modifications)

Element name	**Carbon**		**Silicon**	**Germanium**	**Tin**	**Lead**		
Chemical symbol	**C**		**Si**	**Ge**	**Sn**	**Pb**		
Atomic number Z	**6**		**14**	**32**	**50**	**82**		
							Units	**Remarks**
Modification	Diamond	Graphite			β-Sn,[a] white tin, Sn I			
Crystal system, Bravais lattice	cub	hex	cub, fc	cub, fc	tetr, I	cub, fc		
Structure type	Diamond	C	Diamond	Diamond	β-Sn	Cu		
Lattice constant a	356.71	246.12	543.102 (22.5 °C)	565.9(1) RT	581.97 (300 K)	495.02	pm	
Lattice constant c		670.90			317.49		pm	
Space group	$Fd3m$	$P6_3/mmc$	$Fd3m$	$Fd3m$	$I4_1/amd$	$Fm3m$		
Schoenflies symbol	O_h^7	D_{6v}^4	O_h^7	O_h^7	D_{4h}^{19}	O_h^5		
Strukturbericht type		A9	A4	A4	A5	A1		
Pearson symbol		hP4	cF8	cF8	tI4	cF4		
Number A of atoms per cell	8	4	8	8	4	4		
Coordination number	4	3	4	4		12		
Shortest interatomic distance, solid	154.45	142.10	235	244	302	349	pm	Graphite, within layers
		335.45					pm	Graphite, between layers
Range of stability	RT, > 60 GPa	RTP						

[a] At ambient pressure Sn crystallizes in the diamond structure (gray tin, α-Sn) and below 17 °C in the β-tin structure (white tin, β-Sn, Sn-I) at room temperature. If it is alloyed with In or Hg, the simple hexagonal γ-Sn structure is observed

Table 2.1-13B(b) Elements of Group IVA (CAS notation), or Group 14 (new IUPAC notation). Mechanical properties

Element name	Carbon		Silicon	Germanium	Tin	Lead		
Chemical symbol	C		Si	Ge	Sn	Pb		
Atomic number Z	6		14	32	50	82		
							Units	Remarks
Modification	Diamond	Graphite			β-Sn (white)			
Density ϱ, solid	3.513	2.266	2.33	5.32	7.30	11.4	g/cm^3	At 293 K
Density ϱ, liquid			2.525	5.500	6.978	10.678	g/cm^3	
Molar volume V_{mol}	3.42	5.3	12.06	13.64	16.24	18.26	cm^3/mol	
Viscosity η, liquid			2.0		2.71	1.67	mPa s	
Surface tension γ, liquid			0.735	0.650	0.545	0.470	N/m	
Temperature coefficient			-0.5×10^{-3}	-0.20×10^{-3}	-0.075×10^{-3}	-0.26×10^{-3}	N/(m K)	
Coefficient of linear thermal expansion α	1.06×10^{-6}	1.9×10^{-6} [a]	2.56×10^{-6}	5.57×10^{-6}	21.2×10^{-6}	29.1×10^{-6}	1/K	At 293 K
Sound velocity, liquid					2270	1790	m/s	At T_m, 12 MHz
Sound velocity, solid, transverse			5845	2420	1650	710	m/s	At 298 K
Sound velocity, solid, longitudinal	11 220	3450	8433	4580	3300	2050	m/s	
Compressibility κ	2.25×10^{-6}	1.56×10^{-6}	10.2×10^{-6}	13.4×10^{-6}	18.3×10^{-6}	23.7×10^{-6}	1/MPa	Volume compressibility
Bulk modulus B_0	444		98.0	74.9	56.6		GPa	At 295 K
Elastic modulus E	545		112	79.9	52.9	15.8	GPa	
Shear modulus G			80.5	29.6	19.9	5.54	GPa	
Poisson number μ				0.34	0.33	0.44		
Elastic compliance s_{11}	0.951	0.98	7.73	9.73	41.6	93.7	1/TPa	
Elastic compliance s_{33}		27.5			14.9		1/TPa	
Elastic compliance s_{44}	1.732	250	12.7	14.9	45.6	68.0	1/TPa	
Elastic compliance s_{66}					42.8		1/TPa	
Elastic compliance s_{12}	−0.0987	−0.16	−2.15	−2.64	31.2	−43.0	1/TPa	
Elastic compliance s_{13}		−0.33			4.6		1/TPa	
Elastic stiffness c_{11}	1079	1060	165.6	129	72.30	48.8	GPa	At 300 K
Elastic stiffness c_{33}		36.5			88.40		GPa	At 300 K
Elastic stiffness c_{44}	578	4	79.6	67.1	22.03	14.8	GPa	At 300 K
Elastic stiffness c_{66}					24.00		GPa	At 300 K
Elastic stiffness c_{12}	124.5	180	63.9	48.3	59.40	41.4	GPa	At 300 K
Elastic stiffness c_{13}		15			35.78		GPa	At 300 K
Tensile strength			690			700	MPa	
Vickers hardness			2350			39		
Mohs hardness	10							

[a] At 298 K, parallel to layer planes. The corresponding value perpendicular to the layer planes is 2.9×10^{-6}/K

Table 2.1–13B(c) Elements of Group IVA (CAS notation), or Group 14 (new IUPAC notation). Thermal and thermodynamic properties

Element name	**Carbon**		**Silicon**	**Germanium**	**Tin**	**Lead**		
Chemical symbol	**C**		**Si**	**Ge**	**Sn**	**Pb**		
Atomic number Z	**6**		**14**	**32**	**50**	**82**		
							Units	**Remarks**
Modification	Diamond	Graphite			β-Sn (white)			
Thermal conductivity λ	1000–2320	5.7[a]	83.7	58.6	66.6	35.2	W/(m K)	At 300 K
		1960[b]					W/(m K)	At 300 K
Molar heat capacity c_p	6.11	8.519	20.00	33.347	27.17	26.51	J/(mol K)	At 298 K
Standard entropy S^0	2.360	5.742	18.810	31.090	51.180	64.800	J/(mol K)	At 298 K and 100 kPa
Enthalpy difference $H_{298} - H_0$	0.5188	1.0540	3.2170	4.6360	6.3230	6.8700	kJ/mol	
Melting temperature T_m		4765.30	1687.00	1211.0	505.08	600.61	K	
Enthalpy change ΔH_m		117.3690	50.208	36.9447	7.01940	4.7739	kJ/mol	
Entropy change ΔS_m		24.630	29.762	30.498	14.243	7.948	J/(mol K)	
Relative volume change ΔV_m			−0.10	−0.054	0.028	0.032		$(V_l - V_s)/V_l$ at T_m
Boiling temperature T_b		3915	3505	3107	2876	2019	K	
Enthalpy change ΔH_b		710.9	383.3	331	295.8	177.58	kJ/mol	
Transformation temperature	1900–2100						K	Transforms to graphite

[a] Perpendicular to layer planes.
[b] Parallel to layer planes.

Table 2.1-13B(d) Elements of Group IVA (CAS notation), or Group 14 (new IUPAC notation). Electronic, electromagnetic, and optical properties

Element name	Carbon		Silicon	Germanium	Tin	Lead		
Chemical symbol	C		Si	Ge	Sn	Pb		
Atomic number Z	6		14	32	50	82		
							Units	Remarks
Modification	Diamond	Graphite			β-Sn(white)			
Characteristics	Very hard insulator	Soft conductor	Hard semicondor	Semicondor	Ductile metal	Soft metal		
Electrical resistivity ρ_s	10^{11}	1.4×10^{-5}		0.45	110×10^{-9}	192×10^{-9}	Ω m	At 293 K
Temperature coefficient					46.5×10^{-4}	42.8×10^{-4}	1/K	
Pressure coefficient					-9.2×10^{-9}	-12.5×10^{-9}	1/hPa	
Electrical resistivity ρ_l				710			nΩ m	
Resistivity ratio				0.071		1.94		ρ_l/ρ_s at T_m
Superconducting critical temperature T_{crit}					3.72	7.2	K	
Superconducting critical field H_{crit}					309	803	Oe	
Hall coefficient R [a]		-487×10^{-10}	-100	0.1	0.041×10^{-10}	0.09×10^{-10}	$m^3/(A s)$	At 300 K
Thermoelectric coefficient		11.06		302.5	0.1	-0.1	μV/K	
Electronic band gap ΔE	5.4		1.107	0.6642			eV	At 300 K
Temperature dependence			-2.3×10^{-4}				eV/K	At 300 K
Electronic work function		4.81	4.95	5.0	4.42	4.25	V	
Thermal work function		4.00	4.1	4.56	4.11	3.83	V	
Electron mobility	1800		1900	3800			$cm^2/(V s)$	
Hole mobility	1400		480	1820			$cm^2/(V s)$	
Dielectric constant ε, static, solid	5.68		11.7[b]	16.0				
Dielectric constant ε, high-frequency, solid	5.9(1)		12.0	16	24			
Molar magnetic susceptibility χ_{mol}, solid [c] (SI)	-74.1×10^{-6}	-75.4×10^{-6}	-39.2×10^{-6}	-146×10^{-6}	-470×10^{-6}	-289×10^{-6}	cm^3/mol	At 295 K
Molar magnetic susceptibility χ_{mol}, solid [c] (cgs)	-5.88×10^{-6}	-6.0×10^{-6}	-3.12×10^{-6}	-11.6×10^{-6}	-37.4×10^{-6}	-23×10^{-6}	cm^3/mol	At 295 K
Mass magnetic susceptibility χ_{mass}, solid (SI)	-6.17×10^{-6}	-6.3×10^{-6}	-1.8×10^{-6}	-1.328×10^{-6}	-3.3×10^{-6}	-1.39×10^{-6}	cm^3/g	At 295 K
Mass magnetic susceptibility χ_{mass}, liquid (SI)					-4.4×10^{-6}		cm^3/g	
Refractive index n, solid	2.4173		4.24	4.00 (25 μm)	1.0	2.01		$\lambda = 589$ nm
Refractive index n, liquid					1.7			

[a] $B = 1.0–1.6$ T.

[b] At 11 K and 1 MHz.

[c] The values for Sn apply to gray Sn.

Table 2.1–13C Elements of Group IVA (CAS notation), or Group 14 (new IUPAC notation). Allotropic and high-pressure modifications

Element	Silicon				Germanium				
Modification	α-Si	β-Si	γ-Si	δ-Si	α-Ge	β-Ge	γ-Ge	δ-Ge	
									Units
Crystal system, Bravais lattice	cub, fc	tetr	cub	hex	cub, fc	tetr	tetr	cub, bc	
Structure type	Diamond	β-Sn		α-La	Diamond	β-Sn		γ-Si	
Lattice constant a	543.06	468.6	636	380	565.74	488.4	593	692	pm
Lattice constant c		258.5		628		269.2	698		pm
Space group	$Fd\bar{3}m$	$I4_1/amd$	$Im3m$	$P6_3/mmc$	$Fd\bar{3}m$	$I4_1/amd$	$P4_32_12$	$Im\bar{3}m$	
Schoenflies symbol	O_h^7	D_{4h}^{19}	O_h^9	D_{6v}^4	O_h^7	D_{4h}^{19}	D_4^8		
Strukturbericht type	A4	A5		A3′	A4	A5			
Pearson symbol	cF8	tI4	cI16	hP4	cF8	tI4	tP12	cI16	
Number A of atoms per cell	8	4		4	8	4	12	16	
Coordination number	4	4+2			4	4+2	4+2+2		
Shortest interatomic distance, solid	235	243			245	253	249		
Range of stability	RTP	> 9.5 GPa	> 16.0 GPa	[a]	RTP	> 12.0 GPa	[b]	> 12.0 GPa	

[a] Decompressed β-Si.
[b] Decompressed β-Ge.

Table 2.1–13C Elements of Group IVA (CAS notation), or Group 14 (new IUPAC notation). Allotropic and high-pressure modifications, cont.

Element	Tin			Lead		
Modification	α-Sn (gray tin)	β-Sn (white tin)	γ-Sn	Pb I	Pb II	
						Units
Crystal system, Bravais lattice	cub, fc	tetr, I	tetr	cub, fc	hex, cp	
Structure type	Diamond	β-Sn	In	Cu	Mg	
Lattice constant a	648.92	583.16	370	495.02	326.5	pm
Lattice constant c		318.15	337		538.7	pm
Space group	$Fd\bar{3}m$	$I4_1/amd$		$Fm\bar{3}m$	$P6_3/mmc$	
Schoenflies symbol	O_h^7	D_{4h}^{19}		O_h^5	D_{6h}^4	
Strukturbericht type	A4	A5		A1	A3	
Pearson symbol	c8	tI4	tI2	cF4	hP2	
Number A of atoms per cell	8	4	2	4	2	
Coordination number	4			12		
Shortest interatomic distance, solid	281	302		349		
Range of stability	< 291 K	RT	> 9 GPa	RTP	> 10.3 GPa	

Table 2.1–13D Elements of Group IVA (CAS notation), or Group 14 (new IUPAC notation). Ionic radii (determined from crystal structures)

Element	Carbon	Silicon	Germanium		Tin	Lead		
Ion	C^{4+}	Si^{4+}	Ge^{2+}	Ge^{4+}	Sn^{4+}	Pb^{2+}	Pb^{4+}	
Coordination number								Units
4	15	26		39	55		65	pm
6	16	40	73	53	69	119	78	pm
8					81	129	94	pm
10						140		
12						149		pm

Table 2.1–14A Elements of Group IVB (CAS notation), or Group 4 (new IUPAC notation). Atomic, ionic, and molecular properties (see Table 2.1-14D for ionic radii)

Element name	Titanium	Zirconium	Hafnium	Rutherfordium		
Chemical symbol	Ti	Zr	Hf	Rf		
Atomic number Z	22	40	72	104		
					Units	Remarks
Characteristics				Radioactive		
Relative atomic mass A (atomic weight)	47.867(1)	91.224(2)	178.49(2)	[261]		
Abundance in lithosphere	4400×10^{-6}	220×10^{-6}	4.5×10^{-6}			Mass ratio
Abundance in sea	1×10^{-9}	3×10^{-11}	7×10^{-12}			Mass ratio
Atomic radius r_{cov}	132	145	144		pm	Covalent radius
Atomic radius r_{met}	147	160	156		pm	Metallic radius, CN = 12
Electron shells	–LMN	–MNO	–NOP	–OPQ		
Electronic ground state	$^3F_2$	$^3F_2$	$^3F_2$	$^3F_2$		
Electronic configuration	$[Ar]3d^24s^2$	$[Kr]4d^25s^2$	$[Xe]4f^{14}5d^26s^2$			
Oxidation states	4+, 3+	4+	4+			
Electron affinity	0.079	0.426	near 0		eV	
Electronegativity χ_A	1.32	1.22	(1.23)			Allred and Rochow
1st ionization energy	6.8282	6.63390	6.82507		eV	
2nd ionization energy	13.5755	13.13	14.9		eV	
3rd ionization energy	27.4917	22.99	23.3		eV	
4th ionization energy	43.2672	34.34	33.33		eV	
Standard electrode	−1.630	−1.553[a]			V	Reaction type $Ti^{2+} + 2e^- = Ti$
potential E^0	−0.368				V	Reaction type $Ti^{3+} + e^- = Ti^{2+}$
	−0.04				V	Reaction type $Ti^{4+} + e^- = Ti^{3+}$

[a] Reaction type $ZrO_2 + 4H^+ + 4e^- = Zr + 2H_2O$.

Table 2.1-14B(a) Elements of Group IVB (CAS notation), or Group 4 (new IUPAC notation). Crystallographic properties (see Table 2.1-14C for allotropic and high-pressure modifications)

Element name	Titanium	Zirconium	Hafnium	Rutherfordium		
Chemical symbol	Ti	Zr	Hf	Rf		
Atomic number Z	22	40	72	104		
					Units	Remarks
Modification	α-Ti	α-Zr	α-Hf			
Crystal system, Bravais lattice	hex	hex	hex			
Structure type	Mg	Mg	Mg			
Lattice constant a	295.03	323.17	319.46		pm	
Lattice constant c	468.36	514.76	505.11		pm	
Space group	$P6_3/mmc$	$P6_3/mmc$	$P6_3/mmc$			
Schoenflies symbol	D_{6h}^4	D_{6h}^4	D_{6h}^4			
Strukturbericht type	A3	A3	A3			
Pearson symbol	hP2	hP2	hP2			
Number A of atoms per cell	2	2	2			
Coordination number	12	6+6	12			
Shortest interatomic distance, solid	291		318		pm	

Table 2.1-14B(b) Elements of Group IVB (CAS notation), or Group 4 (new IUPAC notation). Mechanical properties

Element name	Titanium	Zirconium	Hafnium	Rutherfordium		
Chemical symbol	Ti	Zr	Hf	Rf		
Atomic number Z	22	40	72	104		
					Units	Remarks
Density ϱ, solid	4.50	6.49	13.10		g/cm^3	
Density ϱ, liquid	4.110	5.80	12.00		g/cm^3	
Molar volume V_{mol}	10.55	14.02	13.41		cm^3/mol	
Surface tension, liquid	1.65	1.48	1.63		N/m	Near T_m
Temperature coefficient	0.26×10^{-3}	-0.2×10^{-3}	-0.21×10^{-3}		N/(m K)	
Coefficient of linear thermal expansion α	8.35×10^{-6}	5.78×10^{-6}	5.9×10^{-6}		1/K	
Sound velocity, solid, transverse	2920	1950	2000		m/s	
Sound velocity, solid, longitudinal	6260	4360	3671		m/s	
Compressibility κ	0.779×10^{-5}	1.08×10^{-5}	0.80×10^{-5}		1/MPa	Volume compressibility
Elastic modulus E	102	68.0	138		GPa	
Shear modulus G	37.3	24.8	53.0		GPa	
Poisson number μ	0.35	0.37	0.29			
Elastic compliance s_{11}	9.69	10.1	7.16		1/TPa	
Elastic compliance s_{33}	6.86	8.0	6.13		1/TPa	
Elastic compliance s_{44}	21.5	30.1	18.0		1/TPa	
Elastic compliance s_{12}	−4.71	−4.0	−2.48		1/TPa	
Elastic compliance s_{13}	−1.82	−2.4	−1.57		1/TPa	
Elastic stiffness c_{11}	160	144	181		GPa	
Elastic stiffness c_{33}	181	166	197		GPa	
Elastic stiffness c_{44}	46.5	33.4	55.7		GPa	
Elastic stiffness c_{12}	90	74	77		GPa	
Elastic stiffness c_{13}	66	67	66		GPa	
Tensile strength	235	150–450	400		MPa	
Vickers hardness	2000–3500	903	1760			

Table 2.1–14B(c) Elements of Group IVB (CAS notation), or Group 4 (new IUPAC notation). Thermal and thermodynamic properties

Element name	Titanium	Zirconium	Hafnium	Rutherfordium		
Chemical symbol	Ti	Zr	Hf	Rf		
Atomic number Z	22	40	72	104		
					Units	Remarks
Modification		α-Zr				
Thermal conductivity λ	21	22.7	23.0		W/(m K)	
Molar heat capacity c_p	25.02	25.36	25.3		J/(mol K)	At 298 K
Standard entropy S^0	30.720	39.181	43.560		J/(mol K)	At 298 K and 100 kPa
Enthalpy difference $H_{298} - H_0$	4.8240	5.5663	5.8450		kJ/mol	
Melting temperature T_m	1941.00	2127.85	2506.00		K	
Enthalpy change ΔH_m	14.1460	20.9978	27.1960		kJ/mol	
Entropy change ΔS_m	7.288	9.868	10.852		J/(mol K)	
Volume change ΔV_m						$(V_l - V_s)/V_l$ at T_m
Boiling temperature T_b	3631	4203	4963		K	
Enthalpy change ΔH_b	410.0	561.3	575.5		kJ/mol	

Table 2.1–14B(d) Elements of Group IVB (CAS notation), or Group 4 (new IUPAC notation). Electronic, electromagnetic, and optical properties

Element name	Titanium	Zirconium	Hafnium	Rutherfordium		
Chemical symbol	Ti	Zr	Hf	Rf		
Atomic number Z	22	40	72	104		
					Units	Remarks
Modification		α-Zr				
Characteristics	Hard, light metal	Resistant metal	Metal			
Electrical resistivity ρ_s	390	410	296		nΩ m	At 293 K
Temperature coefficient	54.6×10^{-4}	44.0×10^{-4}	44×10^{-4}		1/K	
Pressure coefficient	-1.118×10^{-9}	0.33×10^{-9}	-0.87×10^{-9}		1/hPa	
Superconducting critical temperature T_{crit}	0.39	0.55	0.35		K	
Superconducting critical field H_{crit}	100	47			Oe	
Hall coefficient R	-1.2×10^{-10}	0.212×10^{-10}	0.43×10^{-10}		$m^3/(A\,s)$	At 300 K, $B = 0.4–2.8$ T
Electronic work function	4.31	4.05	3.9		V	
Thermal work function	4.16	4.12	3.53		V	
Molar magnetic susceptibility χ_{mol}, solid (SI)	1898×10^{-6}	1508×10^{-6}	892×10^{-6}		cm^3/mol	At 295 K
Molar magnetic susceptibility χ_{mol}, solid (cgs)	151×10^{-6}	120×10^{-6}	71×10^{-6}		cm^3/mol	At 295 K
Mass magnetic susceptibility χ_{mass}, solid (SI)	40.1×10^{-6}	16.8×10^{-6}	5.3×10^{-6}		cm^3/g	At 295 K

Table 2.1–14C Elements of Group IVB (CAS notation), or Group 4 (new IUPAC notation). Allotropic and high-pressure modifications

Element	Titanium		Zirconium		Hafnium			
Modification	α-Ti	β-Ti	α-Zr	β-Zr	α-Hf	β-Hf	Hf-II	
								Units
Crystal system, Bravais lattice	hex, cp	cub, bc	hex, cp	cub, bc	hex, cp	cub, bc	hex	
Structure type	Mg	W	Mg	W	Mg	W		
Lattice constant *a*	295.03	330.65	323.17	360.9	319.46	361.0		pm
Lattice constant *c*	468.36		514.76		505.11			pm
Space group	$P6_3/mmc$	$Im\bar{3}m$	$P6_3/mmc$	$Im\bar{3}m$	$P6_3/mmc$	$Im\bar{3}m$		
Schoenflies symbol	D_{6h}^4	O_h^9	D_{6h}^4	O_h^9	D_{6h}^4	O_h^9		
Strukturbericht type	A3	A2	A3	A2	A3	A2		
Pearson symbol	hP2	cI2	hP2	cI2	hP2	cI2		
Number *A* of atoms per cell	2	2	2	2	2	2		
Coordination number	12	8	6+6	8	12	8		
Shortest interatomic distance, solid	291	286		313	318	313		
Range of stability	RT	> 1173 K	RT	> 1138 K	RTP	> 2268 K	> 38.8 GPa	

Table 2.1–14D Elements of Group IVB (CAS notation), or Group 4 (new IUPAC notation). Ionic radii (determined from crystal structures)

Element	Titanium			Zirconium	Hafnium	
Ion	Ti^{2+}	Ti^{3+}	Ti^{4+}	Zr^{4+}	Hf^{4+}	
Coordination number						Units
4			42	59	58	pm
6	86	67	61	72	71	pm
8			74	84	83	pm
9				89		pm

2.1.5.3 Elements of the Main Groups and Subgroup V to VIII

Table 2.1-15A Elements of Group VA (CAS notation), or Group 15 (new IUPAC notation). Atomic, ionic, and molecular properties (see Table 2.1-15D for ionic radii)

Element name	Nitrogen	Phosphorus	Arsenic	Antimony	Bismuth		
Chemical symbol	N	P	As	Sb	Bi		
Atomic number Z	7	15	33	51	83		
						Units	Remarks
Relative atomic mass A (atomic weight)	14.00674(7)	30.973761(2)	74.92160(2)	121.760(1)	208.98038(2)		
Abundance in lithosphere	20×10^{-6}	1200×10^{-6}	5×10^{-6}	0.2×10^{-6}	0.2×10^{-6}		Mass ratio
Abundance in sea	150×10^{-6}	0.06×10^{-6}	3.7×10^{-9}	2.4×10^{-10}	2×10^{-11}		Mass ratio
Atomic radius r_{cov}	70	110	121	141	146	pm	Covalent radius
Atomic radius r_{met}	92	128	139	159	182	pm	Metallic radius, CN = 12
Atomic radius r_{vdW}	155	190	200	220	240	pm	van der Waals radius
Electron shells	KL	KLM	–LMN	–MNO	–NOP		
Electronic ground state	$^4S_{3/2}$	$^4S_{3/2}$	$^4S_{3/2}$	$^4S_{3/2}$	$^4S_{3/2}$		
Electronic configuration	$[He]2s^22p^3$	$[Ne]3s^23p^3$	$[Ar]3d^{10}4s^24p^3$	$[Kr]4d^{10}5s^25p^3$	$[Xe]4f^{14}5d^{10}6s^26p^3$		
Oxidation states	5+, 3+, 3−	5+, 3+, 3−	5+, 3+, 3−	5+, 3+, 3−	5+, 3+		
Electron affinity	Not stable	0.747	0.81	1.046	0.946	eV	
Electronegativity χ_A	3.07	2.06	2.20	1.82	(1.67)		Allred and Rochow
1st ionization energy	14.53414	10.48669	9.8152	8.64	7.289	eV	
2nd ionization energy	29.6013	19.7694	18.633	16.53051	16.69	eV	
3rd ionization energy	47.44924	30.2027	28.351	25.3	25.56	eV	
4th ionization energy	77.4735	51.4439	50.13	44.2	45.3	eV	
5th ionization energy	97.8902	65.0251	62.63	56	56.0	eV	
6th ionization energy	552.0718	220.421	127.6	108	88.3	eV	

Table 2.1-15B(a) Elements of Group VA (CAS notation), or Group 15 (new IUPAC notation). Crystallographic properties (see Table 2.1-15C for allotropic and high-pressure modifications)

Element name	**Nitrogen**	**Phosphorus**	**Arsenic**	**Antimony**	**Bismuth**		
Chemical symbol	**N**	**P**	**As**	**Sb**	**Bi**		
Atomic number *Z*	**7**	**15**	**33**	**51**	**83**		
						Units	**Remarks**
State	N_2, 4.2 K	P, black					
Crystal system, Bravais lattice	cub, sc	orth, C	trig, R	trig, R	trig, R		
Structure type	α-N	P, black	α-As	α-As	α-As		
Lattice constant a	565.9	331.36	413.20	450.65	474.60	pm	
Lattice constant b		1047.8				pm	
Lattice constant c		43 763				pm	
Lattice angle α			54.12	57.11	57.23	deg	
Space group	$Pa3$	$Cmca$	$R\bar{3}m$	$R\bar{3}m$	$R\bar{3}m$		
Schoenflies symbol	T_h^6	D_{2h}^{18}	D_{3d}^5	D_{3d}^5	D_{3d}^5		
Strukturbericht type		A11	A7	A7	A7		
Pearson symbol	cP8	oC8	hR2	hR2	hR2		
Number A of atoms per cell	4×2	8	2	2	2		
Coordination number		$2+1$	3	3	3		
Shortest interatomic distance, solid		222	252	291	307	pm	

Table 2.1–15B(b) Elements of Group VA (CAS notation), or Group 15 (new IUPAC notation). Mechanical properties

Element name	Nitrogen	Phosphorus	Arsenic	Antimony	Bismuth		
Chemical symbol	**N**	**P**	**As**	**Sb**	**Bi**		
Atomic number Z	**7**	**15**	**33**	**51**	**83**		
						Units	**Remarks**
State	N_2 gas	P, black					
Density ϱ, solid		2.690	5.72	6.68	9.80	g/cm^3	
Density ϱ, liquid					10.05	g/cm^3	
Density ϱ, gas	1.2506×10^{-3}					g/cm^3	At 273 K
Molar volume V_{mol}	13.65		12.95	18.20	21.44	cm^3/mol	
Viscosity η, gas	18.9					µPa s	At 323 K
Viscosity η, liquid				1.50	1.65	mPa s	Near T_m
Surface tension γ, liquid				0.384	0.376	N/m	
Coefficient of linear thermal expansion α			4.7×10^{-6}	8.5×10^{-6}	13.4×10^{-6}	1/K	
Sound velocity, gas	336.9					m/s	
Sound velocity, liquid	929 (70 K)				1635	m/s	
Sound velocity, solid, transverse				1800	1140	m/s	
Sound velocity, solid, longitudinal				3140	2298	m/s	
Compressibility κ				2.6×10^{-5}	2.86×10^{-5}	1/MPa	Volume compressibility
Elastic modulus E	1.2 [a,b]	30.4 [b]	22	54.4	34.0	GPa	
Shear modulus G				20.6	12.8	GPa	
Poisson number μ				0.25	0.33		
Elastic compliance s_{11}			46.71	16.1	26.0	1/TPa	
Elastic compliance s_{33}			202.9	29.9	42.0	1/TPa	
Elastic compliance s_{44}			44.91	38.9	114	1/TPa	
Elastic compliance s_{12}			36.94	−6.1	−7.9	1/TPa	
Elastic compliance s_{13}			−88.2	−6.2	−11.9	1/TPa	
Elastic compliance s_{14}			1.80	−12.3	−21.2	1/TPa	

[a] Solid.

[b] Estimated.

Table 2.1–15B(b) Elements of Group VA (CAS notation), or Group 15 (new IUPAC notation). Mechanical properties, cont.

Element name	**Nitrogen**	**Phosphorus**	**Arsenic**	**Antimony**	**Bismuth**		
Chemical symbol	**N**	**P**	**As**	**Sb**	**Bi**		
Atomic number Z	**7**	**15**	**33**	**51**	**83**		
						Units	**Remarks**
State	N_2 gas	P, black					
Elastic stiffness c_{11}		73.9	123.6	101	63.4	GPa	
Elastic stiffness c_{22}		277				GPa	
Elastic stiffness c_{33}	53.7	53.7	59.1	44.9	37.9	GPa	
Elastic stiffness c_{44}		15.6	22.6	39.5	11.5	GPa	
Elastic stiffness c_{55}		11.5				GPa	
Elastic stiffness c_{66}		56.7				GPa	
Elastic stiffness c_{12}			19.7	32.2	24.5	GPa	
Elastic stiffness c_{13}			62.3	27.6	24.9	GPa	
Elastic stiffness c_{14}			−4.16	21.8	7.3	GPa	
Brinell hardness					200		
Mohs hardness			3.5				
Solubility in water α_W, 293 K	0.01557						α_W = vol (gas)/vol (water)

Table 2.1-15B(c) Elements of Group VA (CAS notation), or Group 15 (new IUPAC notation). Thermal and thermodynamic properties

Element name	Nitrogen	Phosphorus	Arsenic	Antimony	Bismuth		
Chemical symbol Atomic number Z	N 7	P 15	As 33	Sb 51	Bi 83		
						Units	Remarks
State	N_2 gas	P, white [a]	α-As				
Thermal conductivity λ	0.02598		50.0	25.9	7.87	W/(m K)	
Molar heat capacity c_p	14.560	23.824	24.65	25.23	25.52	J/(mol K)	At 298 K
Standard entropy S^0	191.611	41.090	35.689	45.522	56.735	J/(mol K)	At 298 K and 100 kPa
Enthalpy difference $H_{298} - H_0$	8.6692	5.3600	5.1170	5.8702	6.4266	kJ/mol	
Melting temperature T_m	63.1458	317.30	1090.00	903.78	544.55	K	
Enthalpy change ΔH_m	0.720	0.6590	24.4429	19.8740	11.2968	kJ/mol	
Entropy change ΔS_m		2.077	22.425	21.990	20.745	J/(mol K)	
Volume change ΔV_m			0.10	−0.008	−0.33		$(V_l - V_s)/V_l$ at T_m
Boiling temperature T_b	77.35	550		1860	1837	K	
Enthalpy change ΔH_b	5.577	51.9	34.8	165.8	174.1	kJ/mol	
Critical temperature T_c	126.25		1089			K	
Critical pressure p_c	3.40		36			MPa	
Critical density ϱ_c	0.311					g/cm^3	
Triple-point temperature T_{tr}	63.14					K	
Triple-point pressure p_{tr}	12.5					kPa	

[a] In Landolt–Börnstein, Group IV, Vol. 19A, Part 1 [1.4], the white form of phosphorus has been chosen as the reference phase for all phosphides because the more stable red form is difficult to characterize.

Table 2.1-15B(d) Elements of Group VA (CAS notation), or Group 15 (new IUPAC notation). Electronic, electromagnetic, and optical properties

Element name	Nitrogen	Phosphorus	Arsenic	Antimony	Bismuth		
Chemical symbol	N	P	As	Sb	Bi		
Atomic number Z	7	15	33	51	83		
						Units	Remarks
State	N_2 gas	P, black					
Characteristics	Gas	Semiconductor	Semiconductor	Semimetal	Brittle metal		
Electrical resistivity ρ_s			260	370	1068	nΩ m	Solid, at RT
Temperature coefficient			42×10^{-4}	51.1×10^{-4}	45.4×10^{-4}	1/K	
Pressure coefficient				6.0×10^{-9}	15.2×10^{-9}	1/hPa	
Electrical resistvity ρ_l				1135	1280	nΩ m	Liquid
Resistivity ratio				0.61	0.43		ρ_l/ρ_s at T_m
Electronic band gap ΔE		0.57	1.14			eV	At 300 K
Temperature coefficient		8×10^{-4}				eV/K	
Electronic work function			4.79	4.56	4.36	V	
Thermal work function			5.71	4.08	4.28	V	
Electron mobility		220				$cm^2/(V\,s)$	
Hole mobility		350				$cm^2/(V\,s)$	
Hall coefficient R			0.45×10^{-7}	0.27×10^{-7}	-6.33×10^{-7}	$m^3/(A\,s)$	$B = 0.4–1.0\,T$
Thermoelectric coefficient				35	−70	μV/K	
Dielectric constant $(\varepsilon-1)$, gas	580×10^{-6}						At 293 K
Dielectric constant ε, liquid	1.45						At 74.8 K
Dielectric constant ε, solid			11.2 (optical)				
Molar magnetic susceptibility χ_{mol}, (SI)	-151×10^{-6}	a,b	-70.4×10^{-6} c	-1244×10^{-6}	-3520×10^{-6}	cm^3/mol	At 295 K
Molar magnetic susceptibility χ_{mol}, (cgs)	-12.0×10^{-6}	a,b	-5.60×10^{-6} c	-99×10^{-6}	-280.1×10^{-6}	cm^3/mol	At 295 K
Mass magnetic susceptibility χ_{mass}, (SI)	-5.4×10^{-6}		-3.9×10^{-6}	-10×10^{-6}	-16.8×10^{-6}	cm^3/g	At 295 K
Refractive index $(n-1)$, gas	297×10^{-6}						589.3 nm
Refractive index n, liquid	1.929 (78 K)						589.3 nm
Refractive index n, solid			3.35 (0.8 μm)				

[a] White P: molar magnetic susceptibility: -26.7×10^{-6} (cgs) and -335×10^{-6} (SI) cm^3/mol.

[b] Red P: molar magnetic susceptibility: -20.8×10^{-6} (cgs) and -261×10^{-6} (SI) cm^3/mol.

[c] Yellow As: molar magnetic susceptibility: -292×10^{-6} (SI) and -23.2×10^{-6} (cgs) cm^3/mol.

Table 2.1–15C Elements of Group VA (CAS notation), or Group 15 (new IUPAC notation). Allotropic and high-pressure modifications

Element	Nitrogen			Arsenic		Antimony				
Modification	α-N	β-N	γ-N	α-As	ε-As	Sb-I	Sb-II	Sb-III	Sb-IV	
										Units
Crystal system, Bravais lattice	cub, P	hex	tetr	trig, R	orth	trig, R	cub	hex, cp	mon	
Structure type	α-N	La		As	α-Ga	α-As		Mg		
Lattice constant a	565.9	404.6	395.7	413.20	362	450.65	299.2	337.6	556	pm
Lattice constant b					1085				404	pm
Lattice constant c		662.9	510.1		448			534.1	422	pm
Lattice angle α				54.12		57.11				deg
Lattice angle β									86.0	deg
Space group	$Pa3$	$P6_3/mmc$	$P4_2/mnm$	$R\bar{3}m$	$Cmca$	$R\bar{3}m$	$Pm3m$	$P6_3/mmc$		
Schoenflies symbol	T_h^6	D_{6v}^4	D_{4h}^{12}	D_{3d}^5	D_{2h}^{18}	D_{3d}^5	O_h^1	D_{6h}^4		
Strukturbericht type		A9		A7	A11	A7				
Pearson symbol	cP8	hP4	tP4	hR2	oC8	hR2	cP1	A3		
Number A of atoms per cell	4×2	4	4	2	8	2	1	hP2		
Coordination number		3		3		3		2		
Shortest interatomic distance, solid				252		291				pm
Range of stability	< 20 K	> 35.6 K	20 K, > 3.3 GPa		> 721 K	RTP	> 5.0 GPa	> 7.5 GPa	14.0 GPa	

Table 2.1–15C Elements of Group VA (CAS notation), or Group 15 (new IUPAC notation). Allotropic and high-pressure modifications, cont.

Element	Bismuth						
Modification	α-Bi	β-Bi	γ-Bi	δ-Bi	ε-Bi	ζ-Bi	
							Units
Crystal system, Bravais lattice	trig, R	mon	mon			cub, bc	
Structure type	α-As					W	
Lattice constant a	474.60		605			3800	pm
Lattice constant b			420				pm
Lattice constant c			465				pm
Lattice angle α	57.23						deg
Space group	$R\bar{3}m$	$C2/m$				$Im\bar{3}m$	
Schoenflies symbol	D_{3d}^5	C_{2h}^3				O_h^9	
Strukturbericht type	A7					A2	
Pearson symbol	hR2	mC4	mP3			cI2	
Number A of atoms per cell	2	4	3			2	
Coordination number	3						
Shortest interatomic distance, solid	307						pm
Range of stability	RTP	> 0.28 GPa	> 3.0 GPa	> 4.3 GPa	> 6.5 GPa	> 9.0 GPa	

Table 2.1-15D Elements of Group VA (CAS notation), or Group 15 (new IUPAC notation). Ionic radii (determined from crystal structures)

Element	Nitrogen		Phosphorus	Arsenic		Antimony		Bismuth		
Ion	N^{3+}	N^{5+}	P^{5+}	As^{3+}	As^{5+}	Sb^{3+}	Sb^{5+}	Bi^{3+}	Bi^{5+}	
Coordination number										Units
4			17		34	76				pm
5								96		pm
6	16	13	38	58	46	76	60	103	76	pm
8								117		pm

Table 2.1-16A Elements of Group VB (CAS notation), or Group 5 (new IUPAC notation). Atomic, ionic, and molecular properties (see Table 2.1-16D for ionic radii)

Element name	Vanadium	Niobium	Tantalum	Dubnium		
Chemical symbol	V	Nb	Ta	Db		
Atomic number Z	23	41	73	105		
					Units	Remarks
Relative atomic mass A (atomic weight)	50.9415(1)	92.90638(2)	180.9479(1)	[262]		
Abundance in lithosphere	150×10^{-6}	20×10^{-6}	2.1×10^{-6}			Mass ratio
Abundance in sea	2.5×10^{-9}	1×10^{-11}	2×10^{-12}			Mass ratio
Atomic radius r_{cov}	122	134	134		pm	Covalent radius
Atomic radius r_{met}	135	147	147		pm	Metallic radius, CN = 12
Electron shells	–LMN	–MNO	–NOP	–OPQ		
Electronic ground state	$^4F_{3/2}$	$^6D_{1/2}$	$^4F_{3/2}$			
Electronic configuration	$[Ar]3d^34s^2$	$[Kr]4d^45s^1$	$[Xe]4f^{14}5d^36s^2$			
Oxidation states	5+, 4+, 3+, 2+	5+, 3+	5+			
Electron affinity	0.525	0.893	0.322		eV	
Electronegativity χ_A	1.45	1.23	(1.33)			Allred and Rochow
1st ionization energy	6.7463	6.75885	7.89		eV	
2nd ionization energy	14.66	14.32			eV	
3rd ionization energy	29.311	25.04			eV	
4th ionization energy	46.709	38.3			eV	
5th ionization energy	65.282	50.55			eV	
Standard electrode	−1.175				V	Reaction type $V^{2+}+2e^-=V$
potential E^0	−0.255				V	Reaction type $V^{3+}+e^-=V^{2+}$

Table 2.1-16B(a) Elements of Group VB (CAS notation), or Group 5 (new IUPAC notation). Crystallographic properties

Element name	Vanadium	Niobium	Tantalum	Dubnium		
Chemical symbol	V	Nb	Ta	Db		
Atomic number Z	23	41	73	105		
					Units	Remarks
Crystal system, Bravais lattice	cub, bc	cub, bc	cub, bc			
Structure type	W	W	W			
Lattice constant a	302.38	330.07	330.31		pm	
Space group	$Im\bar{3}m$	$Im\bar{3}m$	$Im\bar{3}m$			
Schoenflies symbol	O_h^9	O_h^9	O_h^9			
Strukturbericht type	A2	A2	A2			
Pearson symbol	cI2	cI2	cI2			
Number A of atoms per cell	2	2	2			
Coordination number	8	8	8			
Shortest interatomic distance, solid	263	285	285		pm	

Table 2.1-16B(b) Elements of Group VB (CAS notation), or Group 5 (new IUPAC notation). Mechanical properties

Element name	Vanadium	Niobium	Tantalum	Dubnium		
Chemical symbol	V	Nb	Ta	Db		
Atomic number Z	23	41	73	105		
					Units	Remarks
Density ϱ, solid	5.80	8.35	16.60		g/cm^3	
Density ϱ, liquid	5.55	7.830	15,00		g/cm^3	
Molar volume V_{mol}	8.34	10.84	10.87		cm^3/mol	
Surface tension, liquid	1.95	2.0	2.15		N/m	
Temperature coefficient	0.3×10^{-3}	-0.24×10^{-3}	-0.25×10^{-3}		N/(m K)	
Coefficient of linear thermal expansion α	8.3×10^{-6}	7.34×10^{-6}	6.64×10^{-6}		1/K	300 K
Sound velocity, solid, transverse	2780	2100	2900		m/s	
Sound velocity, solid, longitudinal	6000	4900	4100		m/s	
Compressibility κ	0.63×10^{-5}	0.56×10^{-5}	0.465×10^{-5}		1/MPa	Volume compressibility
Elastic modulus E	127	104	185		GPa	
Shear modulus G	46.6	59.5	64.7		GPa	
Poisson number μ	0.36	0.38	0.35			
Elastic compliance s_{11}	6.75	6.56	6.89		1/TPa	
Elastic compliance s_{44}	23.2	35.2	12.1		1/TPa	
Elastic compliance s_{12}	−2.31	−2.29	−2.58		1/TPa	
Elastic stiffness c_{11}	230	245	264		GPa	
Elastic stiffness c_{44}	43.1	28.4	82.6		GPa	
Elastic stiffness c_{12}	120	132	158		GPa	
Vickers hardness	630	70–250 (HV 10)	80–300 (HV 10)			At 293 K

Table 2.1-16B(c) Elements of Group VB (CAS notation), or Group 5 (new IUPAC notation). Thermal and thermodynamic properties

Element name	Vanadium	Niobium	Tantalum	Dubnium		
Chemical symbol	**V**	**Nb**	**Ta**	**Db**		
Atomic number Z	**23**	**41**	**73**	**105**		
					Units	**Remarks**
Thermal conductivity λ	30.7	59.0	60.7		W/(m K)	At 300 K
Molar heat capacity c_p	24.90	24.69	25.30		J/(mol K)	At 298 K
Standard entropy S^0	30.890	36.270	41.472		J/(mol K)	At 298 K and 100 kPa
Enthalpy difference $H_{298} - H_0$	4.5070	5.2200	5.6819		kJ/mol	
Melting temperature T_m	2183.00	2750.00	3290.00		K	
Enthalpy change ΔH_m	21.5000	30.0000	36.5682		kJ/mol	
Entropy change ΔS_m	9.849	10.909	11.115		J/(mol K)	
Relative volume change ΔV_m						$(V_l - V_s)/V_l$ at T_m
Boiling temperature T_b	3690	5017	5778		K	
Enthalpy change ΔH_b	451.8	683.2	743.1		kJ/mol	

Table 2.1-16B(d) Elements of Group VB (CAS notation), or Group 5 (new IUPAC notation). Electronic, electromagnetic, and optical properties. There is no Table 2.1-16C, because no allotropic or high-pressure modifictions are known

Element name	Vanadium	Niobium	Tantalum	Dubnium		
Chemical symbol	**V**	**Nb**	**Ta**	**Db**		
Atomic number Z	**23**	**41**	**73**	**105**		
					Units	**Remarks**
Electrical resistivity ρ_s	248	152	125		nΩ m	At RT
Temperature coefficient	39.0×10^{-4}	25.8×10^{-4}	38.2×10^{-4}		1/K	
Pressure coefficient	-1.6×10^{-9}	-1.37×10^{-9}	-1.62×10^{-9}		1/hPa	
Superconducting critical temperature T_{crit}	5.3	9.13	4.49		K	
Superconducting critical field H_{crit}	1020	1980	830		Oe	
Hall coefficient R	0.82×10^{-10}	0.88×10^{-10}	1.01×10^{-10}		$m^3/(A\,s)$	At 273 K, $B = 0.5–2.9$ T
Thermoelectric coefficient			−5.0		μV/K	
Electronic work function	4.3	4.3	4.3		V	
Thermal work function	4.09	3.99	4.25		V	
Molar magnetic susceptibility, solid χ_{mol}, (SI)	3581×10^{-6}	2614×10^{-6}	1935×10^{-6}		cm^3/mol	At 295 K
Molar magnetic susceptibility, solid χ_{mol}, (cgs)	285×10^{-6}	208×10^{-6}	154×10^{-6}		cm^3/mol	At 295 K
Mass magnetic susceptibility, solid χ_{mass}, (SI)	62.8×10^{-6}	27.6×10^{-6}	10.7×10^{-6}		cm^3/g	At 295 K

Table 2.1-16D Elements of Group VB (CAS notation), or Group 5 (new IUPAC notation). Ionic radii (determined from crystal structures)

Element	Vanadium				Niobium			Tantalum			
Ion	V^{2+}	V^{3+}	V^{4+}	V^{5+}	Nb^{3+}	Nb^{4+}	Nb^{5+}	Ta^{3+}	Ta^{4+}	Ta^{5+}	
Coordination number											**Units**
4				36			48				pm
5			53	46							pm
6	79	64	58	54	72	68	64	72	68	64	pm
8			72		79		74				pm

Table 2.1-17A Elements of Group VIA (CAS notation), or Group 16 (new IUPAP notation). Atomic, ionic and molecular properties (see Table 2.1-17D for ionic radii)

Element name	Oxygen	Sulfur	Selenium	Tellurium	Polonium		
Chemical symbol **Atomic number** Z	**O** **8**	**S** **16**	**Se** **34**	**Te** **52**	**Po** **84**		
						Units	**Remarks**
Characteristics					Radioactive		
Relative atomic mass A (atomic weight)	15.9994(3)	32.066(6)	78.96(3)	127.60(3)	[209]		
Abundance in lithosphere	$464\,000 \times 10^{-6}$	520×10^{-6}	0.09×10^{-6}	2×10^{-9}			Mass ratio
Abundance in sea	$880\,000 \times 10^{-6}$	905×10^{-6}	2×10^{-10}				Mass ratio
Atomic radius r_{cov}	66	104	117	137	146	pm	Covalent radius
Atomic radius r_{met}		127	140	160	176	pm	Metallic radius, CN = 12
Atomic radius r_{vdW}	150	185	200	220			van der Waals radius
Electron shells	KL	KLM	–LMN	–MNO	–NOP		
Electronic ground state	$^3P_2$	$^3P_2$	$^3P_2$	$^3P_2$	$^3P_2$		
Electronic configuration	$[He]2s^22p^4$	$[Ne]3s^23p^4$	$[Ar]3d^{10}4s^24p^4$	$[Kr]4d^{10}5s^25p^4$	$[Xe]4f^{14}5d^{10}6s^26p^4$		
Oxidation states	2–	6+, 4+, 2+, 2–	6+, 4+, 2+, 2–	6+, 4+, 2+, 2–	4+, 2+		
Electron affinity	1.46	2.08	2.02	1.97	1.9	eV	Reaction type $O + e^- = O^-$
	–8.75	–5.51				eV	Reaction type $O^- + e^- = O^{2-}$
Electronegativity χ_A	3.50	2.44	2.48	2.01	(1.76)		Allred and Rochow
1st ionization energy	13.61806	10.36001	9.75238	9.0096	8.41671	eV	
2nd ionization energy	35.11730	23.3379	21.19	18.6		eV	
3rd ionization energy	54.9355	34.79	30.8204	27.96		eV	
4th ionization energy	77.41353	47.222	42.9450	37.41		eV	
Standard electrode potential E^0		–0.476	–0.924	–1.143		V	Reaction type $Te^{2-} = Te + 2e^-$
					+0.56	V	Reaction type $Po^{3+} + 3e^- = Po$
				+0.568		V	Reaction type $Te^{4+} + 4e^- = Te$

Table 2.1-17B(a) Elements of Group VIA (CAS notation), or Group 16 (new IUPAC notation). Crystallographic properties (see Table 2.1-17C for allotropic and high-pressure modifications)

Element name	**Oxygen**	**Sulfur**	**Selenium**	**Tellurium**	**Polonium**		
Chemical symbol	**O**	**S**	**Se**	**Te**	**Po**		
Atomic number *Z*	**8**	**16**	**34**	**52**	**84**		
						Units	**Remarks**
State	α-O, $T < 23$ K	S_8, α-S	Gray Se, Se chains		α-Po		
Crystal system, Bravais lattice	mon, C	orth, F	hex	hex	cub, sc		
Structure type	α-O	α-S	γ-Se	γ-Se	α-Po		
Lattice constant a	540.3	1046.4	436.55	445.61	336.6	pm	
Lattice constant b	342.9	1286.60				pm	
Lattice constant c	508.6	2448.60	495.76	592.71		pm	
Lattice angle γ	132.53					deg	
Space group	$C2/m$	$Fddd$	$P3_121$	$P3_121$	$Pm3m$		
Schoenflies symbol	C_{2h}^3	D_{2h}^{24}	D_3^4	D_3^4	O_h^1		
Strukturbericht type		A16	A8	A8	A_h		
Pearson symbol	mC4	oF128	hP3	hP3	cP1		
Number A of atoms per cell	2×2	128	3	3	1		
Coordination number		2	2	2+4	6		
Shortest interatomic distance, solid		204	237	283	337	pm	

Table 2.1-17B(b) Elements of Group VIA (CAS notation), or Group 16 (new IUPAC notation). Mechanical properties

Element name	Oxygen	Sulfur	Selenium	Tellurium	Polonium		
Chemical symbol	O	S	Se	Te	Po		
Atomic number Z	8	16	34	52	84		
						Units	Remarks
State	O_2 gas	α-S	Gray Se				
Density ϱ, solid		2.037	4.79	6.24	9.40	g/cm^3	At 293 K
Density ϱ, liquid		1.819	3.990	5.797		g/cm^3	
Density ϱ, gas	1.429×10^{-3}					g/cm^3	
Molar volume V_{mol}	8.00	15.49	16.48	20.45	22.4	cm^3/mol	
Viscosity η, gas	19.5					μPa s	
Viscosity η, liquid		11.5	1260			mPa s	At T_m
Surface tension, liquid		0.061	0.106	0.186		N/m	At T_m
Coefficient of linear thermal expansion α		74.33×10^{-6}	36.9×10^{-6}	16.75×10^{-6}	23×10^{-6}	1/K	
Sound velocity, gas	336.95 (70 K)					m/s	
Sound velocity, liquid	1079 (70 K)					m/s	
Compressibility κ		13.0×10^{-5}	11.6×10^{-5}	4.8×10^{-5}		1/MPa	Volume compressibilty
Elastic modulus E		17.8 [a]	58.0	47.1	26 [a]	GPa	
Shear modulus G			6.46	16.7		GPa	
Poisson number μ			0.45	0.23			
Elastic compliance s_{11}	1280 [b]	74.6				1/TPa	
Elastic compliance s_{22}		111				1/TPa	
Elastic compliance s_{33}		75.4				1/TPa	
Elastic compliance s_{44}	3640 [b]	121				1/TPa	
Elastic compliance s_{55}		234				1/TPa	
Elastic compliance s_{66}		229				1/TPa	
Elastic compliance s_{12}	−570 [b]	−13.1				1/TPa	
Elastic compliance s_{13}		−7.1				1/TPa	
Elastic compliance s_{23}		−45.8				1/TPa	
Elastic stiffness c_{11}	2.60 [b]	14.22				GPa	
Elastic stiffness c_{22}		12.68				GPa	
Elastic stiffness c_{33}		18.30				GPa	
Elastic stiffness c_{44}	0.275 [b]	8.27				GPa	
Elastic stiffness c_{55}		4.28				GPa	
Elastic stiffness c_{66}		4.37				GPa	
Elastic stiffness c_{12}	2.06 [b]	2.99				GPa	
Elastic stiffness c_{13}		3.14				GPa	
Elastic stiffness c_{23}		7.95				GPa	
Tensile strength				10.8–12.25		MPa	
Brinell hardness				250			
Mohs hardness			2.0				
Solubility in water α_W [c]	0.0310						At 293 K and 1013 hPa

[a] Estimated. [b] γ-Oxygen, $T = 54.4$ K. [c] α_W = vol (gas)/vol (water).

Table 2.1-17B(c) Elements of Group VIA (CAS notation), or Group 16 (new IUPAC notation). Thermal and thermodynamic properties

Element name	**Oxygen**	**Sulfur**	**Selenium**	**Tellurium**	**Polonium**		
Chemical symbol	**O**	**S**	**Se**	**Te**	**Po**		
Atomic number Z	**8**	**16**	**34**	**52**	**84**		
						Units	**Remarks**
State	O_2 gas	α-S	α-Se	α-Te	α-Po		
Thermal conductivity λ	0.0245	0.269	2.48	1.7	20	W/(m K)	At STP
Molar heat capacity c_p	14.690	22.70	25.04	25.73		J/(mol K)	At 298 K
Standard entropy S^0	205.147	32.070	41.966	49.221	62.000	J/(mol K)	At 298 K and 100 kPa
Enthalpy difference $H_{298} - H_0$	8.6800	4.4120	5.5145	6.0800	6.700	kJ/mol	
Melting temperature T_m	54.361	388.36	494	722.66	527.00	K	
Enthalpy change ΔH_m	0.444	1.7210	6.6944	17.3760	10.000	kJ/mol	
Entropy change ΔS_m		4.431	13.551	24.045	18.975	J/(mol K)	
Relative volume change ΔV_m		0.515	+0.168	0.05			$(V_l - V_s)/V_l$ at T_m
Boiling temperature T_b	90.18	882	958	1261	1335	K	
Enthalpy change ΔH_b	6.2	9.62	90	104.6	100.8	kJ/mol	
Critical temperature T_c	154.58		1863			K	
Critical pressure p_c	5.4		38			MPa	
Critical density ϱ_c	0.419					g/cm^3	
Triple-point temperature T_{tr}	54.4					K	
Triple-point pressure p_{tr}	1.52					hPa	

Table 2.1-17B(d) Elements of Group VIA (CAS notation), or Group 16 (new IUPAC notation). Electronic, electromagnetic, and optical properties

Element name	Oxygen	Sulfur	Selenium	Tellurium	Polonium		
Chemical symbol	O	S	Se	Te	Po		
Atomic number Z	8	16	34	52	84		
						Units	Remarks
State	O_2 gas		Gray Se				
Characteristics		Solid insulator	Semiconductor	Semiconductor	Volatile metal		
Electrical resistivity ρ_s			100	1–50		MΩ m	At RT
Electrical resistivity ρ_l			20	6.0 μΩ m		MΩ m	
Resistivity ratio at T_m			1.0	0.048–0.091			
Hall coefficient R				0.24×10^{-10}		$m^3/(A\,s)$	At 298 K
Thermoelectric coefficient				400		μV/K	
Electronic band gap ΔE		3.6	1.79	0.33		eV	
Temperature coefficient		-6.8×10^{-4}	-9×10^{-4}			eV/K	
Electronic work function			5.9	4.95		V	
Thermal work function			4.72	4.73		V	
Electron mobility				1100		$cm^2/(V\,s)$	
Hole mobility				560		$cm^2/(V\,s)$	
Dielectric constant $(\varepsilon - 1)$, gas	525×10^{-6}						At 373 K
Dielectric constant ε, liquid	1.505						At 81 K
Dielectric constant ε, solid			8.5 ($\lambda = 3.3$ cm)	$5.0 \parallel c$ $2.2 \perp c$			
Molar magnetic susceptibility χ_{mol}(SI)	$43\,341 \times 10^{-6}$ [a]	-195×10^{-6}	-314×10^{-6}	-478×10^{-6}		cm^3/mol	At 295 K
Molar magnetic susceptibility χ_{mol} (cgs)	3449×10^{-6} [b]	-15.5×10^{-6}	-25×10^{-6}	-38×10^{-6}		cm^3/mol	At 295 K
Mass magnetic susceptibility χ_{mass} (SI)	1.34×10^{-6} [c]	-6.09×10^{-6}	-4.0×10^{-6}	-3.9×10^{-6}		cm^3/g	At 295 K
Mass magnetic susceptibility χ_{mass}, liquid (SI)				-0.6×10^{-6}		cm^3/g	
Refractive index $(n - 1)$, gas	270.6×10^{-6}						$\lambda = 589.3$ nm
Refractive index n, liquid	1.221 (92 K)						$\lambda = 589.3$ nm
Refractive index n, solid			4.0				$\lambda = 589.3$ nm

[a] Liquid O_2, 90 K, 96 748 cm^3/mol; solid O_2, 54 K, 128 177 cm^3/mol. [b] Liquid O_2, 90 K, 7699 cm^3/mol; solid O_2, 54 K, 10 200 cm^3/mol.
[c] At 280 K.

Further remarks

Liquid O_3 has a molar magnetic susceptibility of 84.2×10^{-6} cm^3/mol (SI) and 6.7×10^{-6} cm^3/mol (cgs).

The values given for the molar magnetic susceptibility χ_{mol} of sulfur apply to rhombic sulfur. The corresponding values for monoclinic sulfur are -187×10^{-6} cm^3/mol (SI) and -14.9×10^{-6} cm^3/mol (cgs).

Table 2.1-17C Elements of Group VIA (CAS notation), or Group 16 (new IUPAC notation). Allotropic and high-pressure modifications

Element	Oxygen			Selenium			
Modification	**α-O**	**β-O**	**γ-O**	**γ-Se**	**α-Se**	**β-Se**	
							Units
Crystal system, Bravais lattice	mon, C	trig, R	cub, P	hex	mon	mon	
Structure type	α-O	α-As	γ-O	γ-Se			
Lattice constant a	540.3	421.0	683	436.55	905.4	1501.8	pm
Lattice constant b	342.9				908.3	1471.3	pm
Lattice constant c	508.6			495.76	233.6	887.9	pm
Lattice angle α		46.27					deg
Lattice angle γ	132.53				90.82	93.6	deg
Space group	$C2/m$	$R\bar{3}m$	$Pm\bar{3}m$	$P3_121$	$P2_1/m$	$P2_1/b$	
Schoenflies symbol	C_{2h}^3	D_{3d}^5	O_h^3	D_3^4	C_{2h}^2	C_{2h}^5	
Strukturbericht type		A7	A15	A8			
Pearson symbol	mC4	hR2	cP16	hP3		mP32	
Number A of atoms per cell	2×2	2	16	3		32	
Coordination number				2	2	2	
Shortest interatomic distance, solid				237	233–235	233–236	
Range of stability	$<$ 23 K	$>$ 23.9 K	$>$ 43.6 K	RT	RT	RT	
Characteristics				Gray, Se chains	Red, Se_8 rings;	Red, Se_8 rings	

Table 2.1-17C Elements of Group VIA, or Group 16. Allotropic and high-pressure modifications, cont.

Element	Tellurium			Polonium		
Modification	**α-Te**	**β-Te**	**γ-Te**	**α-Po**	**β-Po**	
						Units
Crystal system, Bravais lattice	hex	trig, R	trig, R	cub, P	trig, R	
Structure type	γ-Se	α-As	α-Hg	α-Po	α-Hg	
Lattice constant a	445.61	469	300.2	336.6	337.3	pm
Lattice constant c	592.71					pm
Lattice angle α		53.30	103.3		98.08	deg
Space group	$P3_121$	$R\bar{3}m$	$R\bar{3}m$	$Pm\bar{3}m$	$R\bar{3}m$	
Schoenflies symbol	D_3^4	D_{3d}^5	D_{3d}^5	O_h^1	D_{3d}^5	
Strukturbericht type	A8	A7	A10	A_h	A10	
Pearson symbol	hP3	hR2	hR1	cP1	hR1	
Number A of atoms per cell	3	2	1	1	1	
Coordination number	2+4			6	6	
Shortest interatomic distance, solid	283			337	337	
Range of stability	RTP	$>$ 2 GPa	$>$ 7.0 GPa	RTP	$>$ 327 K	

Table 2.1-17D Elements of Group VIA (CAS notation), or Group 16 (new IUPAC notation). Ionic radii (determined from crystal structures)

Element	Oxygen	Sulfur			Selenium			Tellurium			Polonium	
Ion	O^{2-}	S^{2-2}	S^{4+}	S^{6+}	Se^{2-}	Se^{4+}	Se^{6+}	Te^{2-}	Te^{4+}	Te^{6+}	Po^{4+}	
Coordination number												**Units**
2	121											pm
4				12			28		66	43		pm
6	140	184	37	29	198	50	42	221	97	56	97	pm
8	142											pm

Table 2.1-18A Elements of Group VIB (CAS notation), or Group 6 (new IUPAC notation). Atomic, ionic, and molecular properties (see Table 2.1-18D for ionic radii)

Element name	**Chromium**	**Molybdenum**	**Tungsten**	**Seaborgium**		
Chemical symbol	**Cr**	**Mo**	**W**	**Sg**		
Atomic number Z	**24**	**42**	**74**	**106**		
					Units	**Remarks**
Characteristics				Radioactive		
Relative atomic mass *A* (atomic weight)	51.9961(6)	95.94(1)	183.84(1)			
Abundance in lithosphere	200×10^{-6}	2.3×10^{-6}	1×10^{-6}			Mass ratio
Abundance in sea	3×10^{-10}	0.01×10^{-6}	1×10^{-10}			Mass ratio
Atomic radius r_{cov}	118	130	130		pm	Covalent radius
Atomic radius r_{met}	129	140	141		pm	Metallic radius, CN = 12
Electron shells	–LMN	–MNO	–NOP	–OPQ		
Electronic ground state	$^7S_3$	$^7S_3$	$^5D_0$			
Electronic configuration	$[Ar]3d^5 4s^1$	$[Kr]4d^5 5s^1$	$[Xe]4f^{14} 5d^4 6s^2$			
Oxidation states	6+, 3+, 2+	6+, 5+, 4+, 3+, 2+, 1+, 2–	6+, 5+, 4+, 3+, 2+, 1–, 2–			
Electron affinity	0.666	0.748	0.815		eV	
Electronegativity χ_A	1.56	1.30	(1.40)			Allred and Rochow
1st ionization energy	6.76664	7.09243	7.98		eV	
2nd ionization energy	16.4857	16.16			eV	
3rd ionization energy	30.96	27.13			eV	
4th ionization energy	49.16	46.4			eV	
Standard electrode	−0.913				V	Reaction type $Cr^{2+} + 2e^- = Cr$
potential E^0	−0.744	−0.2			V	Reaction type $Cr^{3+} + 3e^- = Cr$
	−0.407				V	Reaction type $Cr^{3+} + e^- = Cr^{2+}$

Table 2.1-18B(a) Elements of Group VIB (CAS notation), or Group 6 (new IUPAC notation). Crystallographic properties (see Table 2.1-18C for allotropic and high-pressure modifications)

Element name	Chromium	Molybdenum	Tungsten	Seaborgium		
Chemical symbol	Cr	Mo	W	Sg		
Atomic number Z	24	42	74	106		
					Units	Remarks
Crystal system, Bravais lattice	cub, bc	cub, bc	cub, bc			
Structure type	W	W	W			
Lattice constant a	288.47	314.70	316.51		pm	
Space group	$Im\bar{3}m$	$Im\bar{3}m$	$Im\bar{3}m$			
Schoenflies symbol	O_h^9	O_h^9	O_h^9			
Strukturbericht type	A2	A2	A2			
Pearson symbol	cI2	cI2	cI2			
Number A of atoms per cell	2	2	2			
Coordination number	8	8	8			
Shortest interatomic distance, solid	249	272	273		pm	

Table 2.1-18B(b) Elements of Group VIB (CAS notation), or Group 6 (new IUPAC notation). Mechanical properties

Element name	Chromium	Molybdenum	Tungsten	Seaborgium		
Chemical symbol	Cr	Mo	W	Sg		
Atomic number Z	24	42	74	106		
					Units	Remarks
Density ϱ, solid	7.19	10.220	19.30		g/cm^3	
Density ϱ, liquid	6.460		17.60		g/cm^3	
Molar volume V_{mol}	7.23	9.39	9.53		cm^3/mol	
Surface tension, liquid	1.6	2.25	2.31		N/m	
Temperature coefficient		-0.3×10^{-3}	-0.29×10^{-3}		N/(m K)	
Coefficient of linear thermal expansion α	6.2×10^{-6}	5.35×10^{-6}	4.31×10^{-6}		1/K	300 K
Sound velocity, solid, transverse	3980	3350	2870		m/s	
Sound velocity, solid, longitudinal	6850	6250	5180		m/s	
Compressibility κ	0.78×10^{-5}	0.338×10^{-5}	0.28×10^{-5}		1/MPa	Volume compressibility
Elastic modulus E	145	330	407		GPa	
Shear modulus G	71.6	123	152		GPa	
Poisson number μ	0.31	0.31	0.28			
Elastic compliance s_{11}	3.05	2.63	2.45		1/TPa	
Elastic compliance s_{44}	9.98	9.20	6.24		1/TPa	
Elastic compliance s_{12}	−0.49	−0.68	−0.69		1/TPa	
Elastic stiffness c_{11}	348	465	523		GPa	
Elastic stiffness c_{44}	100.0	109	160		GPa	
Elastic stiffness c_{12}	67	163	203		GPa	
Tensile strength	Strongly dependent on microstructure					
Vickers hardness	1060	160–400 (HV 10)	360–600 (HV 30)			

Table 2.1-18B(c) Elements of Group VIB (CAS notation), or Group 6 (new IUPAC notation). Thermal and thermodynamic properties

Element name	Chromium	Molybdenum	Tungsten	Seaborgium		
Chemical symbol	Cr	Mo	W	Sg		
Atomic number Z	24	42	74	106		
					Units	Remarks
Thermal conductivity λ	93.7	142	164		W/(m K)	At 293 K
Molar heat capacity c_p	23.44	23.932	24.27		J/(mol K)	At 298 K
Standard entropy S^0	23.543	28.560	32.618		J/(mol K)	At 298 K and 100 kPa
Enthalpy difference $H_{298} - H_0$	4.0500	4.5890	4.9700		kJ/mol	
Melting temperature T_m	2180.00	2893	3693		K	
Enthalpy change ΔH_m	21.0040	37.4798	52.3137		kJ/mol	
Entropy change ΔS_m	9.635	12.942	14.158		J/(mol K)	
Relative volume change ΔV_m						$(V_l - V_s)/V_l$ at T_m
Boiling temperature T_b	2952	4952	5828		K	
Enthalpy chnage ΔH_b	344.3	582.2	806.8		kJ/mol	
Critical temperature T_c		11 000			K	
Critical pressure p_c		540			MPa	
Critical density ϱ_c		2.630			g/cm^3	

Table 2.1-18B(d) Elements of Group VIB (CAS notation), or Group 6 (new IUPAC notation). Electronic, electromagnetic, and optical properties

Element name	Chromium	Molybdenum	Tungsten	Seaborgium		
Chemical symbol	Cr	Mo	W	Sg		
Atomic number Z	24	42	74	106		
					Units	Remarks
Electrical resistivity ρ_s	127	52	54.9		nΩ m	At 293 K
Temperature coefficient	30.1×10^{-4}	43.3×10^{-4}	51×10^{-4}		1/K	
Pressure coefficient	-17.3×10^{-9}	-1.29×10^{-9}	-1.333×10^{-9}		1/hPa	
Superconducting critical temperature T_{crit}		0.92	0.005		K	
Superconducting critical field H_{crit}		98			Oe	
Hall coefficient R	3.63×10^{-10}	1.26×10^{-10}	0.856×10^{-10}		m^3/(A s)	$B = 0.5$–2.0 T, $T = 293$ K
Thermoelectric coefficient		5.9	1.5		μV/K	
Electronic work function	4.5	4.39	4.54		eV	
Thermal work function	4.6	4.26	4.50		V	
Molar magnetic susceptibility χ_{mol} (SI)	2099×10^{-6}	905×10^{-6}	666×10^{-6}		cm^3/mol	At 295 K
Molar magnetic susceptibility χ_{mol} (cgs)	167×10^{-6}	72×10^{-6}	53×10^{-6}		cm^3/mol	At 295 K
Mass magnetic susceptibility χ_{mass} (SI)	44×10^{-6}	12×10^{-6}	4.0×10^{-6}		cm^3/g	At 295 K

Table 2.1-18C Elements of Group VIB (CAS notation), or Group 6 (new IUPAC notation). Allotropic and high-pressure modifications

Element	**Chromium**		
Modification	**α-Cr**	**α′-Cr**	
			Units
Crystal system, Bravais lattice	cub, bc	cub, bc	
Structure type	W	W	
Lattice constant *a*	288.47	288.2	pm
Space group	$Im\bar{3}m$	$Im\bar{3}m$	
Schoenflies symbol	O_h^9	O_h^9	
Strukturbericht type	A2	A2	
Pearson symbol	cI2	cI2	
Number *A* of atoms per cell	2	2	
Coordination number	8	8	
Shortest interatomic distance, solid	249		
Range of stability	RTP	High pressure	

Table 2.1-18D Elements of Group VIB (CAS notation), or Group 6 (new IUPAC notation). Ionic radii (determined from crystal structures)

Element	**Chromium**				**Molybdenum**				**Tungsten**			
Ion	Cr^{2+}	Cr^{3+}	Cr^{4+}	Cr^{6+}	Mo^{3+}	Mo^{4+}	Mo^{5+}	Mo^{6+}	W^{4+}	W^{5+}	W^{6+}	
Coordination number												**Units**
4			41	26			46	41			42	pm
5											51	pm
6	73	62	55	44	69	65	61	59	66	62	60	pm
7								73				pm

Table 2.1-19A Elements of Group VIIA (CAS notation), or Group 17 (new IUPAC notation). Atomic, ionic, and molecular properties (see Table 2.1-19D for ionic radii)

Element name	Fluorine	Chlorine	Bromine	Iodine	Astatine		
Chemical symbol	F	Cl	Br	I	At		
Atomic number Z	9	17	35	53	85		
						Units	Remarks
Characterization					Radioactive		
Relative atomic mass A (atomic weight)	18.9984032(5)	35.4527(9)	79.904(1)	126.90447(3)	[210]		
Abundance in lithosphere	800×10^{-6}	480×10^{-6}	2.5×10^{-6}	0.3×10^{-6}			Mass ratio
Abundance in sea	1.3×10^{-6}	$18\,800\times10^{-6}$	67×10^{-6}	0.06×10^{-6}			Mass ratio
Atomic radius r_{cov}	64	99	114	133	145	pm	Covalent radius
Atomic radius r_{vdW}	150–160	175	200	210		pm	van der Waals radius
Electron shells	KL	KLM	–LMN	–MNO	–NOP		
Electronic ground state	$^2P_{3/2}$	$^2P_{3/2}$	$^2P_{3/2}$	$^2P_{3/2}$	$^2P_{3/2}$		
Electronic configuration	$[He]2s^22p^5$	$[Ne]3s^23p^5$	$[Ar]3d^{10}4s^24p^5$	$[Kr]4d^{10}5s^25p^5$	$[Xe]4f^{14}5d^{10}6s^26p^5$		
Oxidation states		1+, 3+, 5+, 7+, 1−, 3−, 5−, 7−					
Electron affinity	3.40	3.61	3.36	3.06	2.8	eV	Reaction type $F+e^-=F^-$
Electronegativity χ_A	4.10	2.83	2.74	2.21	(1.90)		Allred and Rochow
1st ionization energy	17.42282	12.96764	11.81381	10.45126		eV	
2nd ionization energy	34.97082	23.814	21.8	19.1313		eV	
3rd ionization energy	62.7084	39.61	36	33		eV	
4th ionization energy	87.1398	53.4652	47.3			eV	
Standard electrode potential E^0	+2.866	+1.358	+1.066	+0.536		V	Reaction type $2Cl^-=Cl_2+2e^-$
Molecular form in gaseous state	F_2	Cl_2	Br_2	I_2			
Internuclear distance in molecule		198.8				pm	
Dissociation energy		2.475		1.5417		eV	Extrapolated to $T=0\,K$

Table 2.1-19B(a) Elements of Group VIIA (CAS notation), or Group 17 (new IUPAC notation). Crystallographic properties (see Table 2.1-19C for allotropic and high-pressure modifications)

Element name	Fluorine	Chlorine	Bromine	Iodine	Astatine		
Chemical symbol	F	Cl	Br	I	At		
Atomic number Z	9	17	35	53	85		
						Units	Remarks
State	F_2, < 45.6 K	Cl_2, 113 K	Br_2, 123 K	I_2			
Crystal system, Bravais lattice	mon, C	orth, C	orth, C	orth, C			
Structure type				Ga			
Lattice constant a	550	624	668	726.8		pm	
Lattice constant b	328	448	449	479.7		pm	
Lattice constant c	728	826	874	979.7		pm	
Lattice angle β	102.17					deg	
Space group	$C2/m$	$Cmca$	$Cmca$	$Cmca$			
Schoenflies symbol	C_{2h}^3	D_{2h}^{18}	D_{2h}^{18}	D_{2h}^{18}			
Strukturbericht type	C34	A11	A11	A11			
Pearson symbol	mC6	oC8	oC8	oC8			
Number A of atoms per cell	6	2×4	2×4	2×4			
Coordination number	1	1	1	1			
Shortest interatomic distance, solid	149	198.0	227	269		pm	

Table 2.1-19B(b) Elements of Group VIIA (CAS notation), or Group 17 (new IUPAC notation). Mechanical properties

Element name	Fluorine	Chlorine	Bromine	Iodine	Astatine		
Chemical symbol	**F**	**Cl**	**Br**	**I**	**At**		
Atomic number Z	**9**	**17**	**35**	**53**	**85**		
						Units	**Remarks**
State	F_2, gas	Cl_2, gas	Br_2, crystalline	I_2, crystalline			
Density ϱ, solid				4.92		g/cm^3	
Density ϱ, liquid			3.119 (293 K)				
Density ϱ, gas	1.696×10^{-3}	3.17×10^{-3}				g/cm^3	At 273 K, 1 bar
Molar volume V_{mol}	18.05	17.46	19.73	25.74		cm^3/mol	
Viscosity η, gas	209.3×10^2					μPa s	
Viscosity η, liquid			0.916 (299 K)	2.27		mPa s	
Surface tension, liquid			0.0441 (286 K)	0.0557		N/m	
Sound velocity, gas	336 (375 K)	206				m/s	STP
Elastic compliance s_{11}				328		1/TPa	
Elastic compliance s_{22}				103		1/TPa	
Elastic compliance s_{33}				132		1/TPa	
Elastic compliance s_{44}				303		1/TPa	
Elastic compliance s_{55}				67.7		1/TPa	
Elastic compliance s_{66}				170		1/TPa	
Elastic compliance s_{12}				−97		1/TPa	
Elastic compliance s_{13}				−173		1/TPa	
Elastic compliance s_{23}				49		1/TPa	
Elastic stiffness c_{11}			11.5			GPa	
Elastic stiffness c_{22}			13.5			GPa	
Elastic stiffness c_{33}			25.0			GPa	
Elastic stiffness c_{44}			3.30			GPa	
Elastic stiffness c_{55}			14.8			GPa	
Elastic stiffness c_{66}			5.88			GPa	
Elastic stiffness c_{12}			4.50			GPa	
Elastic stiffness c_{13}			13.5			GPa	
Elastic stiffness c_{23}			0.93			GPa	

Table 2.1-19B(c) Elements of Group VIIA (CAS notation), or Group 17 (new IUPAC notation). Thermal and thermodynamic properties

Element name	Fluorine	Chlorine	Bromine	Iodine	Astatine		
Chemical symbol	F	Cl	Br	I	At		
Atomic number Z	9	17	35	53	85		
						Units	Remarks
State	F_2, gas	Cl_2, gas	Br_2, liquid	I_2, crystalline	At_2, crystalline		
Thermal conductivity λ	2.43×10^{-2}	9.3×10^{-3}		0.4	1.7	W/(m K)	STP
Molar heat capacity c_p	15.66	16.974	37.84	27.21		J/(mol K)	At 298 K
Standard entropy S^0	202.789	223.079	152.210	116.139	54.000	J / (mol K)	At 298 K and 100 kPa
Enthalpy difference $H_{298} - H_0$	8.8250	9.1810		13.1963	13.4000	kJ/mol	
Melting temperature T_m	53.48	172.18	265.90	386.75	575	K	
Enthalpy change ΔH_m		6.41	10.8	15.5172	23.8	kJ/mol	
Entropy change ΔS_m				40.122	40.0	J/(mol K)	
Volume change ΔV_m							$(V_l - V_s)/V_l$ at T_m
Boiling temperature T_b	84.95	239.1	332.3	458.4		K	
Enthalpy change ΔH_b		20.40	29.56	41.96	91	kJ/mol	
Critical temperature T_c	144.3	417		819		K	
Critical pressure p_c	5.22	7.98				MPa	
Critical density ϱ_c	0.630	0.573				g/cm^3	
Triple-point temperature T_{tr}	55	162				K	
Triple-point pressure p_{tr}	0.221	1.39				kPa	
Solubility in water[a]		4.610					At 273 K and 101 kPa

[a] m^3 gas/m^3 water.

Table 2.1-19B(d) Elements of Group VIIA (CAS notation), or Group 17 (new IUPAC notation). Electronic, electromagnetic, and optical properties

Element name	Fluorine	Chlorine	Bromine	Iodine	Astatine		
Chemical symbol	**F**	**Cl**	**Br**	**I**	**At**		
Atomic number Z	**9**	**17**	**35**	**53**	**85**		
						Units	**Remarks**
State	F_2, gas	Cl_2, gas	Br_2, liquid	I_2, crystalline			
Characteristics	Yellow gas, very reactive	Yellow-green gas	Liquid halogen	Solid semiconductor			
Electronic work function				2.8		eV	
Dielectric constant ε, liquid	1.517(83.2 K)	2.15 (213 K)					
Molar magnetic susceptibility χ_{mol}, gas (SI)			−924			cm^3/mol	
Molar magnetic susceptibility χ_{mol}, gas (cgs)			-73.5×10^{-6}			cm^3/mol	
Molar magnetic susceptibility χ_{mol}, liquid (SI)		-508×10^{-6}	−709			cm^3/mol	
Molar magnetic susceptibility χ_{mol}, liquid (cgs)		-40.4×10^{-6}	-56.4×10^{-6}			cm^3/mol	
Molar magnetic susceptibility χ_{mol}, solid (SI)				-1131×10^{-6}		cm^3/mol	
Molar magnetic susceptibility χ_{mol}, solid (cgs)				-90×10^{-6}		cm^3/mol	
Mass magnetic susceptibility χ_{mass}, solid (SI)		-7.2×10^{-6}	-11.1×10^{-6}	-4.40×10^{-6}		cm^3/g	
Refractive index $(n-1)$, gas [a]	206×10^{-6}	773×10^{-6}					At 273 K and 101 kPa
Refractive index n, liquid [a]		1.367 [b]	1.659				

[a] $\lambda = 589$ nm. [b] At 92 K, $\varrho = 1.330\ g/cm^3$.

Table 2.1-19C Elements of Group VIIA (CAS notation), or Group 17 (new IUPAC notation). Allotropic and high-pressure modifications

Element	**Fluorine**		
Modification	**α-F (F_2)**	**β-F**	
			Units
Crystal system, Bravais lattice	mon, C	cub, P	
Structure type		γ-O	
Lattice constant *a*	550	667	pm
Lattice constant *b*	328		pm
Lattice constant *c*	728		pm
Lattice angle γ	102.17		deg
Space group	$C2/m$	$Pm\bar{3}m$	
Schoenflies symbol	C_{2h}^3	O_h^3	
Strukturbericht type	C34	A15	
Pearson symbol	mC6	cP16	
Number *A* of atoms per cell	6	16	
Coordination number	1		
Shortest interatomic distance, solid	149		pm
Range of stability	< 45.6 K	> 45.6 K	

Table 2.1-19D Elements of Group VIIA (CAS notation), or Group 17 (new IUPAC notation). Ionic radii (determined from crystal structures)

Element	**Fluorine**		**Chlorine**			**Bromine**			**Iodine**			
Ion	F^-	F^{7+}	Cl^-	Cl^{5+}	Cl^{7+}	Br^-	Br^{5+}	Br^{7+}	I^-	I^{5+}	I^{7+}	
Coordination number												**Units**
3				12			31			44		pm
3					8			25			42	pm
6	133	8	181			196		39	220	95	53	pm

Table 2.1-20A Elements of Group VIIB (CAS notation), or Group 7 (new IUPAC notation). Atomic, ionic, and molecular properties (see Table 2.1-20D for ionic radii)

Element name	**Manganese**	**Technetium**	**Rhenium**	**Bohrium**		
Chemical symbol **Atomic number *Z***	**Mn** **25**	**Tc** **43**	**Re** **75**	**Bh** **107**		
					Units	**Remarks**
Characteristics		Radioactive				
Relative atomic mass *A* (atomic weight)	54.938049(9)	[98]	186.207(1)			
Abundance in lithosphere	1000×10^{-6}		1×10^{-9}			Mass ratio
Abundance in sea	2×10^{-10}					Mass ratio
Atomic radius r_{cov}	118	127	128		pm	Covalent radius
Atomic radius r_{met}	137	137	137		pm	Metallic radius, CN = 12
Electron shells	–LMN	–MNO	–NOP	–OPQ		
Electronic ground state	$^6S_{5/2}$	$^6S_{5/2}$	$^6S_{5/2}$			
Electronic configuration	$[Ar]3d^5 4s^2$	$[Kr]4d^5 5s^2$	$[Xe]4f^{14} 5d^5 6s^2$			
Oxidation states	7+, 6+, 4+ 3+, 2+	7+	7+, 6+, 4+, 2+, 1−			
Electron affinity	Not stable	0.55	0.15		eV	Reaction type $Tc + e^- = Tc^-$
Electronegativity χ_A	1.60	1.36	(1.46)			Allred and Rochow
1st ionization energy	7.43402	7.28	7.88		eV	
2nd ionization energy	15.63999	15.26			eV	
3rd ionization energy	33.668	29.54			eV	
4th ionization energy	51.2				eV	
Standard electrode potential E^0	−1.185				V	Reaction type $Mn^{2+} + 2e^- = Mn$

Table 2.1-20B(a) Elements of Group VIIB (CAS notation), or Group 7 (new IUPAC notation). Crystallographic properties (see Table 2.1-20C for allotropic and high-pressure modifications)

Element name	Manganese	Technetium	Rhenium	Bohrium		
Chemical symbol	Mn	Tc	Re	Bh		
Atomic number Z	25	43	75	107		
					Units	Remarks
State	α-Mn					
Crystal system, Bravais lattice	cub, bc	hex	hex			
Structure type	α-Mn	Mg	Mg			
Lattice constant a	892.19	273.8	276.08		pm	
Lattice constant c		439.4	445.80		pm	
Space group	$Im\bar{3}m$	$P6_3/mmc$	$P6_3/mmc$			
Schoenflies symbol	T_d^3	D_{6h}^4	D_{6h}^4			
Strukturbericht type	A12	A3	A3			
Pearson symbol	cI58	hP2	hP2			
Number A of atoms per cell	58	2	2			
Coordination number		12	12			
Shortest interatomic distance, solid	226–293	274			pm	

Table 2.1-20B(b) Elements of Group VIIB (CAS notation), or Group 7 (new IUPAC notation). Mechanical properties

Element name	Manganese	Technetium	Rhenium	Bohrium		
Chemical symbol	Mn	Tc	Re	Bh		
Atomic number Z	25	43	75	107		
					Units	Remarks
Density ϱ, solid	7.470	11.50	21.00		g/cm^3	
Density ϱ, liquid	6.430		18.80		g/cm^3	
Molar volume V_{mol}	7.38	8.6	8.86		cm^3/mol	
Surface tension, liquid	1.10		2.65		N/m	
Temperature coefficient	0		-0.34×10^{-3}		N/(m K)	
Coefficient of linear thermal expansion α	23×10^{-6}	8.06×10^{-6}	6.63×10^{-6}		1/K	
Sound velocity, solid, transverse	3280	50.6	2930		m/s	
Sound velocity, solid, longitudinal	5560	3270	5360		m/s	
Compressibility κ	0.716×10^{-5}	3.22×10^{-5}	0.264×10^{-5}		1/MPa	Volume compressibility
Elastic modulus E	196	407	520		GPa	At 293 K
Shear modulus G	79.4	162	180		GPa	
Poisson number μ	0.24	0.26	0.26			
Elastic compliance s_{11}		3.2	2.11		1/TPa	
Elastic compliance s_{33}		2.9	1.70		1/TPa	
Elastic compliance s_{44}		5.7	6.21		1/TPa	
Elastic compliance s_{12}		−1.1	−0.80		1/TPa	
Elastic compliance s_{13}		−0.9	−0.40		1/TPa	
Elastic stiffness c_{11}		433	616		GPa	
Elastic stiffness c_{33}		470	683		GPa	
Elastic stiffness c_{44}		177	161		GPa	
Elastic stiffness c_{12}		199	273		GPa	
Elastic stiffness c_{13}		199	206		GPa	
Tensile strength		0.40–0.74	1.16		GPa	
Vickers hardness	9.81	1.510	2.45–8.00			

Table 2.1–20B(c) Elements of Group VIIB (CAS notation), or Group 7 (new IUPAC notation). Thermal and thermodynamic properties

Element name	Manganese	Technetium	Rhenium	Bohrium		
Chemical symbol	Mn	Tc	Re	Bh		
Atomic number Z	25	43	75	107		
					Units	Remarks
State	α-Mn					
Thermal conductivity λ	29.7	185	71.2		W/(m K)	
Molar heat capacity c_p	26.28		25.31		J/(mol K)	At 298 K
Standard entropy S^0	32.220	32.985	36.482		J/(mol K)	At 298 K and 100 kPa
Enthalpy difference $H_{298} - H_0$	4.9957		5.3330		kJ/mol	
Melting temperature T_m	1519.00	2430.01	3458.00		K	
Transition	δ–liquid	α–liquid	α–liquid			
Enthalpy change ΔH_m	12.9089	33.2912	34.0750		kJ/mol	
Entropy change ΔS_m	8.498	13.700	9.854		J/(mol K)	
Relative volume change ΔV_m	0.017					$(V_l - V_s)/V_l$ at T_m
Boiling temperature T_b	2335	4538	5869		K	
Enthalpy change ΔH_b	226.7	592.9	714.8		kJ/mol	
Critical temperature T_c			2090		K	
Critical pressure p_c			14.5		MPa	
Critical density ϱ_c			0.320		g/cm^3	

Table 2.1–20B(d) Elements of Group VIIB (CAS notation), or Group 7 (new IUPAC notation). Electronic, electromagnetic, and optical properties

Element name	Manganese	Technetium	Rhenium	Bohrium		
Chemical symbol	Mn	Tc	Re	Bh		
Atomic number Z	25	43	75	107		
					Units	Remarks
Characteristics	Brittle metal	Metal	Metal			
Electrical resistivity ρ_s	1380	1510	172		nΩ m	At RT
Temperature coefficient	5.0×10^{-4}		44.8×10^{-4}		1/K	
Pressure coefficient	-3.54×10^{-9}				1/hPa	
Electrical resistivity ρ_l	400				nΩ m	
Resistivity ratio at T_m	0.61					
Superconducting critical temperature T_{crit}			1.70		K	
Superconducting critical field H_{crit}			198		Oe	
Hall coefficient R	0.84×10^{-10}		3.15×10^{-10}		m^3/(A s)	At 297 K, $B = 0.5$–5.0 T
Electronic work function	4.08		About 5.0		V	
Thermal work function	3.91		4.96		V	
Molar magnetic susceptibility χ_{mol} (SI)	6421×10^{-6}	1445×10^{-6}	842×10^{-6}		cm^3/mol	At 295 K
Molar magnetic susceptibility χ_{mol} (cgs)	511×10^{-6}	115×10^{-6}	67×10^{-6}		cm^3/mol	At 295 K
Mass magnetic susceptibility χ_{mass} (SI)	121×10^{-6}	31×10^{-6}	4.56×10^{-6}		cm^3/g	At 295 K

Table 2.1-20C Elements of Group VIIB (CAS notation), or Group 7 (new IUPAC notation). Allotropic and high-pressure modifications

Element	Manganese				
Modification	**α-Mn**	**β-Mn**	**γ-Mn**	**δ-Mn**	
					Units
Crystal system, Bravais lattice	cub, bc	cub, P	cub, fc	cub, bc	
Structure type	α-Mn	β-Mn	Cu	W	
Lattice constant *a*	892.19	631.52	386.24	308.06	pm
Lattice constant *c*					pm
Space group	$Im\bar{3}m$	$P4_132$	$Fm\bar{3}m$	$Im\bar{3}m$	
Schoenflies symbol	T_d^3	O'	O_h^5	O_h^9	
Strukturbericht type	A12	A13	A1	A2	
Pearson symbol	cI58	cP20	cF4	cI2	
Number *A* of atoms per cell	58	20	4	2	
Coordination number			12	8	
Shortest interatomic distance, solid	226–293		273	267	pm
Range of stability	RTP	> 1000 K	> 1368 K	> 1408 K	

Table 2.1-20D Elements of Group VIIB (CAS notation), or Group 7 (new IUPAC notation). Ionic radii (determined from crystal structures)

Element	Manganese						Technetium	Rhenium				
Ion	Mn^{2+}	Mn^{3+}	Mn^{4+}	Mn^{5+}	Mn^{6+}	Mn^{7+}	Tc^{4+}	Re^{4+}	Re^{5+}	Re^{6+}	Re^{7+}	
Coordination number												**Units**
4	66		39	33	26	25					38	pm
6	83	58	53				65	63	58	55		pm
8	96											pm

Table 2.1-21A Elements of Group VIIIA (CAS notation), or Group 18 (new IUPAC notation). Atomic, ionic, and molecular properties

Element name	Neon	Argon	Krypton	Xenon	Radon		
Chemical symbol	Ne	Ar	Kr	Xe	Rn		
Atomic number Z	10	18	36	54	86		
						Units	Remarks
Characteristics					Radioactive		
Relative atomic mass A (atomic weight)	20.1797(6)	39.948(1)	83.80(1)	131.29(2)	[222]		
Atomic radius r_{cov}	(69)	(97)	110	130		pm	Covalent radius
Atomic radius r_{met}	154	188	202	216	240	pm	Metallic radius, CN = 12
Atomic radius r_{vdW}	160	188	200	220	240	pm	van der Waals radius
Electron shells	KL	KLM	–LMN	–MNO	–NOP		
Electronic ground state	$^1S_0$	$^1S_0$	$^1S_0$	$^1S_0$	$^1S_0$		
Electronic configuration	$[He]2s^22p^6$	$[Ne]3s^23p^6$	$[Ar]3d^{10}4s^24p^6$	$[Kr]4d^{10}5s^25p^6$	$[Xe]4f^{14}5d^{10}6s^26p^6$		
Electron affinity	Not stable	Not stable	Not stable	Not stable	Not stable	eV	Reaction type $He + e^- = He^-$
Electronegativity χ_A	5.10	3.30	3.10	2.40	(2.06)		Allred and Rochow
1st ionization energy	21.56454	15.75962	13.99961	12.12987	10.74850	eV	
2nd ionization energy	40.96328	27.62967	24.35985	21.20979		eV	
3rd ionization energy	63.45	40.74	36.950	32.1230		eV	
4th ionization energy	97.12	59.81	52.5			eV	

Table 2.1-21B(a) Elements of Group VIIIA (CAS notation), or Group 18 (new IUPAC notation). Crystallographic properties (see Table 2.1-21C for allotropic and high-pressure modifications)

Element name	Neon	Argon	Krypton	Xenon	Radon		
Chemical symbol	Ne	Ar	Kr	Xe	Rn		
Atomic number Z	10	18	36	54	86		
						Units	Remarks
State	At 4.2 K	At 4.2 K	At 4.2 K	At 4.2 K			
Crystal system, Bravais lattice	cub, fc	cub, fc	cub, fc	cub, fc			
Structure type	Cu	Cu	Cu	Cu			
Lattice constant a	446.22	531.2	564.59	613.2		pm	
Space group	$Fm\bar{3}m$	$Fm\bar{3}m$	$Fm\bar{3}m$	$Fm\bar{3}m$			
Schoenflies symbol	O_h^5	O_h^5	O_h^5	O_h^5			
Strukturbericht type	A1	A1	A1	A1			
Pearson symbol	cF4	cF4	cF4	cF4			
Number A of atoms per cell	4	4	4	4			
Coordination number	12	12	12	12			
Shortest interatomic distance, solid	320	383	405	441		pm	

Table 2.1-21B(b) Elements of Group VIIIA (CAS notation), or Group 18 (new IUPAC notation). Mechanical properties

Element name	Neon	Argon	Krypton	Xenon	Radon		
Chemical symbol	Ne	Ar	Kr	Xe	Rn		
Atomic number Z	10	18	36	54	86		
						Units	Remarks
Characterics	Noble gas	Noble gas	Noble gas	Noble gas	Noble gas		
Density ϱ, solid		1.736 (40 K)				g/cm^3	
Density ϱ, gas	0.8994×10^{-3}		3.7493×10^{-3}	5.8971×10^{-3}	9.73×10^{-3}	g/cm^3	At 273 K
Molar volume V_{mol}	13.97		29.68	37.09	50.5	cm^3/mol	
Viscosity η, gas	29.8	21	23.4	21.2		μPa s	At 293 K
Sound velocity, gas	461	308	213	168		m/s	At STP
Sound velocity, liquid		855 (85 K)				m/s	
Elastic modulus E	1.0	1.6	1.8			GPa	Low temperature, estimated values
Elastic compliance s_{11}	1020 (4.7 K)	593 (80 K)	618 (115 K)	690 (160.5 K)		1/TPa	
Elastic compliance s_{44}	1000 (4.7 K)	1073 (80 K)	744 (115 K)	708 (160.5 K)		1/TPa	
Elastic compliance s_{12}	−370 (4.7 K)	−205 (80 K)	−226 (115 K)	−271 (160.5 K)		1/TPa	
Elastic stiffness c_{11}	1.69 (4.7 K)	2.77 (80 K)	2.85 (115 K)	2.93 (160.5 K)		GPa	
Elastic stiffness c_{44}	1.00 (4.7 K)	0.98 (80 K)	1.35 (115 K)	1.41 (160.5 K)		GPa	
Elastic stiffness c_{12}	0.97 (4.7 K)	1.37 (80 K)	1.60 (115 K)	1.89 (160.5 K)		GPa	
Solubility in water α_W	0.010	0.0340	0.059	0.108			At 293 K, α_W = vol(gas)/vol(water)

Table 2.1-21B(c) Elements of Group VIIIA (CAS notation), or Group 18 (new IUPAC notation). Thermal and thermodynamic properties

Element name	Neon	Argon	Krypton	Xenon	Radon		
Chemical symbol	Ne	Ar	Kr	Xe	Rn		
Atomic number Z	10	18	36	54	86		
						Units	Remarks
State	Ne gas	Ar gas	Kr gas	Xe gas	Rn gas		
Thermal conductivity λ	49.3×10^{-3}	18.0×10^{-3}	9.49×10^{-3}	5.1×10^{-3}	3.64×10^{-3}	W/(m K)	
Molar heat capacity c_p	20.786	20.87	20.786	20.744	20.786	J/(mol K)	At 298.15 K
Standard entropy S^0	146.328	154.842	164.085	169.575	176.234	J/(mol K)	At 298.15 K and 100 kPa
Enthalpy difference $H_{298} - H_0$	6.1965	6.1965	6.1965	6.1970	6.1970	kJ/mol	At 298.15 K
Melting temperature T_m	24.563	83.8	115.765	161.391	202	K	
Transition	α–liquid	α–liquid		α–liquid	α–liquid		
Enthalpy change ΔH_m	27.0	1.21	1.64	3.10	2.7	kJ/mol	
Boiling temperature T_b	27.10	87.30	119.80	165.03	211	K	
Enthalpy change ΔH_b	0.324	6.3	9.05	12.65	18.1	kJ/mol	
Critical temperature T_c	44.0	150.75	209.4	289.74		K	
Critical pressure p_c	2.75	4.86	5.50	5.840		MPa	
Critical density ϱ_c	0.4835	0.307	0.9085	1.105		g/cm^3	
Triple-point temperature T_{tr}	24.5	83.85	115.95	161.25		K	
Triple-point pressure p_{tr}	43.3	68.75	73.19	81.6		kPa	

Table 2.1-21B(d) Elements of Group VIIIA (CAS notation), or Group 18 (new IUPAC notation). Electronic, electromagnetic, and optical properties

Element name	Neon	Argon	Krypton	Xenon	Radon		
Chemical symbol	Ne	Ar	Kr	Xe	Rn		
Atomic number Z	10	18	36	54	86		
						Units	Remarks
State	Ne gas	Ar gas	Kr gas	Xe gas	Rn gas		
Dielectric constant $\varepsilon - 1$, gas	130×10^{-6}	545×10^{-6}	7×10^{-6}	1238×10^{-6}			At STP
Dielectric constant ε, liquid		1.516 (89 K)					
Molar magnetic susceptibility χ_{mol}, gas (SI)	-87.5×10^{-6}	-243×10^{-6}	-364×10^{-6}	-572×10^{-6}		cm^3/mol	At 295 K
Molar magnetic susceptibility χ_{mol}, gas (cgs)	-6.96×10^{-6}	-19.3×10^{-6}	-29.0×10^{-6}	-45.5×10^{-6}		cm^3/mol	At 295 K
Mass magnetic susceptibility χ_{mass}, gas (SI)	-4.2×10^{-6}	-6.16×10^{-6}	-4.32×10^{-6}	-4.20×10^{-6}		cm^3/g	At 293 K
Refractive index $(n-1)$, gas	7.25×10^{-6}	281×10^{-6}		706×10^{-6}			$\lambda = 589.3$ nm
Refractive index n, liquid		1.233 (84 K)					$\lambda = 589.3$ nm

Table 2.1-21C Elements of Group VIIIA (CAS notation), or Group 18 (new IUPAC notation). Allotropic and high-pressure modifications. There is no Table 2.1-21D, because no data on ionic radii are available

Element	Argon		
Modification	α-Ar	β-Ar	
			Units
Crystal system, Bravais lattice	cub, fc	hex, cp	
Structure type	Cu	Mg	
Lattice constant a	531.2	376.0	pm
Lattice constant c		614.1	pm
Space group	$Fm\bar{3}m$	$P6_3/mmc$	
Schoenflies symbol	O_h^5	D_{6h}^4	
Strukturbericht type	A1	A3	
Pearson symbol	cF4	hP2	
Number A of atoms per cell	4	2	
Coordination number	12	12	
Shortest interatomic distance, solid	383		pm
Range of stability	< 83.8 K	> 83.8 K	

Table 2.1-22A Elements of Group VIII(1) (CAS notation), or Group 8 (new IUPAC notation). Atomic, ionic, and molecular properties (see Table 2.1-22D for ionic radii)

Element name	Iron	Ruthenium	Osmium	Hassium		
Chemical symbol	Fe	Ru	Os	Hs		
Atomic number Z	26	44	76	108		
Quantity					**Units**	**Remarks**
Relative atomic mass A (atomic weight)	55.845(2)	101.07(2)	190.23(3)			
Abundance in lithosphere	$50\,000 \times 10^{-6}$					Mass ratio
Abundance in sea	2×10^{-9}					Mass ratio
Atomic radius r_{cov}	116	125	126		pm	Covalent radius
Atomic radius r_{met}	126	132.5	134		pm	Metallic radius, CN = 12
Electron shells	–LMN	–MNO	–NOP	–OPQ		
Electronic ground state	$^5D_4$	$^5F_5$	$^5D_4$			
Electronic configuration	$[Ar]3d^64s^2$	$[Kr]4d^75s^1$	$[Xe]4f^{14}5d^66s^2$			
Oxidation states	3+, 2+	8+, 6+, 4+, 3+, 2+	8+, 6+, 4+, 3+, 2+			
Electron affinity	0.151	1.05	1.1		eV	Reaction type $Fe + e^- = Fe^-$
Electronegativity χ_A	1.64	1.42	(1.52)			Allred and Rochow
1st ionization energy	7.9024	7.36050	8.7		eV	
2nd ionization energy	16.1878	16.76			eV	
3rd ionization energy	30.652	28.47			eV	
4th ionization energy	54.8				eV	
5th ionization energy	75.0				eV	
Standard electrode potential E^0	−0.447				V	Reaction type $Fe^{2+} + 2e^- = Fe$
	−0.037				V	Reaction type $Fe^{3+} + 3e^- = Fe$
	+0.771				V	Reaction type $Fe^{3+} + e^- = Fe^{2+}$

Table 2.1-22B(a) Elements of Group VIII(1) (CAS notation), or Group 8 (new IUPAC notation). Crystallographic properties (see Table 2.1-22C for allotropic and high-pressure modifications)

Element name	Iron	Ruthenium	Osmium	Hassium		
Chemical symbol	Fe	Ru	Os	Hs		
Atomic number Z	26	44	76	108		
					Units	Remarks
Modification	α-Fe					
Crystal system, Bravais lattice	cub, bc	hex	hex			
Structure type	W	Mg	Mg			
Lattice constant a	286.65	270.53	273.48		pm	
Lattice constant c		428.14	439.13		pm	
Space group	$Im\bar{3}m$	$P6_3/mmc$	$P6_3/mmc$			
Schoenflies symbol	O_h^9	D_{6h}^4	D_{6h}^4			
Strukturbericht type	A2	A3	A3			
Pearson symbol	cI2	hP2	hP2			
Number A of atoms per cell	2	2	2			
Coordination number	8	6+6	6+6			
Shortest interatomic distance, solid	248	265	267		pm	

Table 2.1-22B(b) Elements of Group VIII(1) (CAS notation), or Group 8 (new IUPAC notation). Mechanical properties

Element name	Iron	Ruthenium	Osmium	Hassium		
Chemical symbol	Fe	Ru	Os	Hs		
Atomic number Z	26	44	76	108		
					Units	Remarks
Density ϱ, solid	7.86	12.20	22.4		g/cm^3	
Density ϱ, liquid	7.020	10.90	20.100		g/cm^3	
Molar volume V_{mol}	7.09	8.14	8.43		cm^3/mol	
Viscosity η, liquid	5.53				mPa s	
Surface tension, liquid	1.65	2.25	2.5		N/m	
Temperature coefficient		-0.31×10^{-3}	-0.33×10^{-3}		N/(m K)	
Coefficient of linear thermal expansion α	12.3×10^{-6}	9.1×10^{-6}	6.1×10^{-6}		1/K	
Sound velocity, solid, transverse	3220	3740	3340		m/s	
Sound velocity, solid, longitudinal	5920	6530	5480		m/s	
Compressibility κ	0.56×10^{-5}	0.331×10^{-5}	0.261×10^{-5}		1/MPa	Volume compressibility
Elastic modulus E	211	432	559		GPa	
Shear modulus G	80.4	173	223		GPa	
Poisson number μ	0.29	0.25	0.25			
Elastic compliance s_{11}	7.67	2.09			1/TPa	
Elastic compliance s_{33}		1.82			1/TPa	
Elastic compliance s_{44}	8.57	5.53			1/TPa	
Elastic compliance s_{12}	−2.83	−0.58			1/TPa	
Elastic compliance s_{13}		−0.41			1/TPa	
Elastic stiffness c_{11}	230	563			GPa	
Elastic stiffness c_{33}		624			GPa	
Elastic stiffness c_{44}	117	181			GPa	
Elastic stiffness c_{12}	135	188			GPa	
Elastic stiffness c_{13}		168			GPa	
Tensile strength	193–206 [a]	540			MPa	
Vickers hardness	608 [a]	$2-5\times10^3$	800			

[a] Strongly dependent on microstructure

Table 2.1-22B(c) Elements of Group VIII(1) (CAS notation), or Group 8 (new IUPAC notation). Thermal and thermodynamic properties

Element name	**Iron**	**Ruthenium**	**Osmium**	**Hassium**		
Chemical symbol	**Fe**	**Ru**	**Os**	**Hs**		
Atomic number Z	**26**	**44**	**76**	**108**		
					Units	**Remarks**
Modification	α-Fe					
Thermal conductivity λ	80.2	117	87.6		W/(m K)	
Molar heat capacity c_p	25.10	24.06	24.7		J/(mol K)	At 298.2 K
Standard entropy S^0	27.280	28.614	32.635		J/(mol K)	At 298.15 K and 100 kPa
Enthalpy difference $H_{298} - H_0$	4.4890	4.6024			kJ/mol	At 298.15 K
Melting temperature T_m	1811.0	2607.0	3306.0		K	
Transition	δ–liquid	α–liquid	δ–liquid			
Enthalpy change ΔH_m	13.8060	38.5890	57.8550		kJ/mol	
Entropy change ΔS_m	7.623	14.802	17.500		J/(mol K)	
Relative volume change ΔV_m	0.034					$(V_l - V_s)/V_l$ at T_m
Boiling temperature T_b	3139	4423	5285		K	
Enthalpy change ΔH_b	349.6	595.5	746		kJ/mol	

Table 2.1-22B(d) Elements of Group VIII(1) (CAS notation), or Group 8 (new IUPAC notation). Electronic, electromagnetic, and optical properties

Element name	**Iron**	**Ruthenium**	**Osmium**	**Hassium**		
Chemical symbol	**Fe**	**Ru**	**Os**	**Hs**		
Atomic number Z	**26**	**44**	**76**	**108**		
					Units	**Remarks**
Characteristics	Soft metal	Metal	Brittle metal			
Electrical resistivity ρ_s	89	76	81		nΩ m	At RT
Temperature coefficient	65.1×10^{-4}	45.8×10^{-4}	42×10^{-4}		1/K	
Pressure coefficient	-2.34×10^{-9}	-2.48×10^{-9}			1/hPa	
Superconducting critical temperature T_{crit}		0.49	0.66		K	
Superconducting critical field H_{crit}		66	65		Oe	
Hall coefficient R	8×10^{-10}	2.2×10^{-10}			$m^3/(A s)$	At 300 K, $B = 4–5$ T
Thermoelectric coefficient	−51.34				µV/K	
Electronic work function	4.70	4.71			V	
Thermal work function	4.50	4.73	4.83		V	
Molar magnetic susceptibility χ_{mol} (SI)	Ferromagnetic	490	138		cm^3/mol	At 295 K
Molar magnetic susceptibility χ_{mol} (cgs)	Ferromagnetic	39	11		cm^3/mol	At 295 K
Mass magnetic susceptibility χ_{mass} (SI)	Ferromagnetic	5.37×10^{-6}	0.65×10^{-6}		cm^3/g	At 295 K

Table 2.1-22C Elements of Group VIII(1) (CAS notation), or Group 8 (new IUPAC notation). Allotropic and high-pressure modifications

Element	Iron				
Modification	α-Fe	γ-Fe	δ-Fe	ε-Fe	
					Units
Crystal system, Bravais lattice	cub, bc	cub, fc	cub, bc	hex, cp	
Structure type	W	Cu	W	Mg	
Lattice constant a	286.65	364.67	291.35	248.5	pm
Lattice constant c				399.0	pm
Space group	$Im\bar{3}m$	$Fm\bar{3}m$	$Im\bar{3}m$	$P6_3/mmc$	
Schoenflies symbol	O_h^9	O_h^5	O_h^9	D_{6h}^4	
Strukturbericht type	A2	A1	A2	A3	
Pearson symbol	cI2	cF4	cI2	hP2	
Number A of atoms per cell	2	4	2	2	
Coordination number	8	12	8	6+6	
Shortest interatomic distance, solid	248	258	254	241	pm
Range of stability	RTP	> 1183 K	> 1663 K	> 13.0 GPa	

Table 2.1-22D Elements of Group VIII(1) (CAS notation), or Group 8 (new IUPAC notation). Ionic radii (determined from crystal structures)

Element	Iron		Ruthenium					Osmium				
Ion	Fe^{2+}	Fe^{3+}	Ru^{3+}	Ru^{4+}	Ru^{5+}	Ru^{7+}	Ru^{8+}	Os^{4+}	Os^{5+}	Os^{6+}	Os^{8+}	
Coordination number												Units
4	63	49				38	36				39	pm
6	61	55	68	62	57			63	58	55		pm
8	92	78										pm

Table 2.1-23A Elements of Group VIII(2) (CAS notation), or Group 9 (new IUPAC notation). Atomic, ionic, and molecular properties (see Table 2.1-23D for ionic radii)

Element name	Cobalt	Rhodium	Iridium	Meitnerium		
Chemical symbol	**Co**	**Rh**	**Ir**	**Mt**		
Atomic number Z	**27**	**45**	**77**	**109**		
					Units	**Remarks**
Relative atomic mass A (atomic weight)	58.933200(9)	102.90550(2)	192.217(3)			
Abundance in lithosphere	40×10^{-6}	0.001×10^{-6}	1×10^{-9}			Mass ratio
Abundance in sea	5×10^{-11}					Mass ratio
Atomic radius r_{cov}	116	125	127		pm	Covalent radius
Atomic radius r_{met}	125	134.5	136		pm	Metallic radius, CN = 12
Electron shells	–LMN	–MNO	–NOP	–OPQ		
Electronic ground state	$^4F_{9/2}$	$^4F_{9/2}$	$^4F_{9/2}$			
Electronic configuration	$[Ar]3d^7 4s^2$	$[Kr]4d^8 5s^1$	$[Xe]4f^{14} 5d^7 6s^2$			
Oxidation states	3+, 2+	4+, 3+, 2+	6+, 4+, 3+, 2+			
Electron affinity	0.662	1.14	1.57		eV	Reaction type $Co + e^- = Co^-$
Electronegativity χ_A	1.75	1.35	(1.44)			Allred and Rochow
1st ionization energy	7.8810	7.45890	9.1		eV	
2nd ionization energy	17.083	18.08			eV	
3rd ionization energy	33.50	31.06			eV	
4th ionization energy	51.3				eV	
Standard electrode potential E^0	−0.28	+0.6			V	Reaction type $Co^{2+} + 2e^- = Co$
			+1.156		V	Reaction type $Ir^{3+} + 3e^- = Ir$

Table 2.1-23B(a) Elements of Group VIII(2) (CAS notation), or Group 9 (new IUPAC notation). Crystallographic properties (see Table 2.1-23C for allotropic and high-pressure modifications)

Element name	Cobalt	Rhodium	Iridium	Meitnerium		
Chemical symbol	**Co**	**Rh**	**Ir**	**Mt**		
Atomic number Z	**27**	**45**	**77**	**109**		
					Units	**Remarks**
Modifaction	ε-Co					
Crystal system, Bravais lattice	hex, cp	cub, fc	cub, fc			
Structure type	Mg	Cu	Cu			
Lattice constant a	250.71	280.32	383.91		pm	
Lattice constant c	406.94				pm	
Space group	$P6_3/mmc$	$Fm\bar{3}m$	$Fm\bar{3}m$			
Schoenflies symbol	D^4_{6h}	O^5_h	O^5_h			
Strukturbericht type	A3	A1	A1			
Pearson symbol	hP2	cF4	cF4			
Number A of atoms per cell	2	4	4			
Coordination number	6+6	12	12			
Shortest interatomic distance, solid	250	269	271		pm	

Table 2.1-23B(b) Elements of Group VIII(2) (CAS notation), or Group 9 (new IUPAC notation). Mechanical properties

Element name	Cobalt	Rhodium	Iridium	Meitnerium		
Chemical symbol	**Co**	**Rh**	**Ir**	**Mt**		
Atomic number Z	**27**	**45**	**77**	**109**		
					Units	**Remarks**
Modification	ε-Co					
Density ϱ, solid	8.90	12.40	22.50		g/cm^3	
Density ϱ, liquid	7.670	10.800	20.000		g/cm^3	
Molar volume V_{mol}	6.62	8.29	8.57		cm^3/mol	
Viscosity η, liquid	4.8				mPa s	Near T_m
Coefficient of linear thermal expansion α	13.36×10^{-6}	8.40×10^{-6}	6.8×10^{-6}		1/K	
Surface tension, liquid	1.520	1.97	2.250		N/m	
Temperature coefficient	-0.92×10^{-3}	-0.3×10^{-3}	-0.31×10^{-3}		N/(m K)	
Sound velocity, solid, transverse	3000	3470	3050		m/s	
Sound velocity, solid, longitudinal	5730	6190	5380		m/s	
Compressibility κ	0.525×10^{-5}	0.350×10^{-5}	0.258×10^{-5}		1/MPa	Volume compressibility
Elastic modulus E	204	379	528		GPa	
Shear modulus G	77.3	149	209		GPa	
Poisson number μ	0.32	0.27	0.26			
Elastic compliance s_{11}	5.11	3.46	2.28		1/TPa	
Elastic compliance s_{33}	3.69				1/TPa	
Elastic compliance s_{44}	14.1	5.43	3.90		1/TPa	
Elastic compliance s_{12}	−2.37	−1.10	−0.67		1/TPa	
Elastic compliance s_{13}	−0.94				1/TPa	
Elastic stiffness c_{11}	295	413	580		GPa	
Elastic stiffness c_{33}	335				GPa	
Elastic stiffness c_{44}	71.0	184	256		GPa	
Elastic stiffness c_{12}	159	194	242		GPa	
Elastic stiffness c_{13}	111				GPa	
Tensile strength	255	951	623×10^3		MPa	
Vickers hardness	1043	1246	1760			

Table 2.1–23B(c) Elements of Group VIII(2) (CAS notation), or Group 9 (new IUPAC notation). Thermal and thermodynamic properties

Element name	Cobalt	Rhodium	Iridium	Meitnerium		
Chemical symbol	Co	Rh	Ir	Mt		
Atomic number Z	27	45	77	109		
					Units	Remarks
Modification	ε-Co					
Thermal conductivity λ	100	150	147		W/(m K)	
Molar heat capacity c_p	24.811	24.98	24.98		J/(mol K)	At 298 K
Standard entropy S^0	30.040	31.556	35.505		J/(mol K)	At 298 K and 100 kPa
Enthalpy difference $H_{298} - H_0$	4.7655		5.2677		kJ/mol	
Melting temperature T_m	1768.0	2237.0	2719.0		K	
Transition	α–liquid	α–liquid	α–liquid			
Enthalpy change ΔH_m	16.200	26.5935	41.124		kJ/mol	
Entropy change ΔS_m	9.163	11.888	15.125		J/(mol K)	
Relative volume change ΔV_m		0.12			1	$(V_l - V_s)/V_l$ at T_m
Boiling temperature T_b	3184	3970	4701		K	
Enthalpy change ΔH_b	376.6	493.3	604.1		kJ/mol	

Table 2.1–23B(d) Elements of Group VIII(2) (CAS notation), or Group 9 (new IUPAC notation). Electronic, electromagnetic, and optical properties

Element name	Cobalt	Rhodium	Iridium	Meitnerium		
Chemical symbol	Co	Rh	Ir	Mt		
Atomic number Z	27	45	77	109		
					Units	Remarks
Modification	α-Co					
Characteristics	Hard metal	Metal	Brittle metal			
Electrical resistivity ρ_s	56	43.0	47		nΩ m	At RT
Temperature coefficient	6.04×10^{-3}	4.62×10^{-3}	4.11×10^{-3}		1/K	
Pressure coefficient	-0.904×10^{-9}	-1.62×10^{-9}	-1.37×10^{-9}		1/hPa	
Electrical resistivity ρ_l	1020				nΩ m	
Resistivity ratio at T_m	1.05					
Superconducting critical temperature T_{crit}			0.14		K	
Superconducting critical field H_{crit}			77		Oe	
Hall coefficient R	360×10^{-12}	50.5×10^{-12}	31.8×10^{-12}		$m^3/(A\,s)$	At 300 K, $B = 4.5–5.0$ T
Thermoelectronic coefficient	17.5	1.0	1.2		μV/K	
Electronic work function	4.97	4.98			V	
Thermal work function	4.37	4.68	5.03		V	
Molar magnetic susceptibility χ_{mol} (SI)	Ferromagnetic	1282×10^{-6}	314×10^{-6}		cm^3/mol	At 295 K
Molar magnetic susceptibility χ_{mol} (cgs)	Ferromagnetic	102×10^{-6}	25×10^{-6}		cm^3/mol	At 295 K
Mass magnetic susceptibility χ_{mass} (SI)	Ferromagnetic	13.6×10^{-6}	1.67×10^{-6}		cm^3/g	At 295 K

Table 2.1-23C Elements of Group VIII(2) (CAS notation), or Group 9 (new IUPAC notation). Allotropic and high-pressure modifications

Element	**Cobalt**		
Modification	**ε-Co**	**α-Co**	
			Units
Crystal system, Bravais lattice	hex, cp	cub, fc	
Structure type	Mg	Cu	
Lattice constant a	250.71	354.45	pm
Lattice constant c	406.94		pm
Space group	$P6_3/mmc$	$Fm\bar{3}m$	
Schoenflies symbol	D_{6h}^4	O_h^5	
Strukturbericht type	A3	A1	
Pearson symbol	hP2	cF4	
Number A of atoms per cell	2	4	
Coordination number	6+6	12	
Shortest interatomic distance, solid	250	251	pm
Range of stability	RTP	> 661 K	

Table 2.1-23D Elements of Group VIII(2) (CAS notation), or Group 9 (new IUPAC notation). Ionic radii (determined from crystal structures)

Element	**Cobalt**		**Rhodium**			**Iridium**			
Ion	Co^{2+}	Co^{3+}	Rh^{3+}	Rh^{4+}	Rh^{5+}	Ir^{3+}	Ir^{4+}	Ir^{5+}	
Coordination number									**Units**
4	56								pm
6	65	55	67	60	55	68	63	57	pm
8	90								pm

Table 2.1-24A Elements of Group VIII(3) (CAS notation) or Group 10 (new IUPAC notation). Atomic, ionic, and molecular properties (see Table 2.1-24D for ionic radii)

Element name	Nickel	Palladium	Platinum			
Chemical symbol Atomic number Z	Ni 28	Pd 46	Pt 78	Uun 110		
					Units	Remarks
Relative atomic mass A (atomic weight)	58.6934(2)	106.42(1)	195.078(2)			
Abundance in lithosphere	100×10^{-6}	0.01×10^{-6}	5×10^{-9}			Mass ratio
Abundance in sea	1.7×10^{-9}					Mass ratio
Atomic radius r_{cov}	115	128	130		pm	Covalent radius
Atomic radius r_{met}	125	138	137		pm	Metallic radius, CN = 12
Atomic radius r_{vdW}	160	160	170–180		pm	van der Waals radius
Electron shells	–LMN	–MNO	–NOP	–OPQ		
Electronic ground state	$^3F_4$	$^1S_0$	$^3D_3$			
Electronic configuration	$[Ar]3d^8 4s^2$	$[Kr]4d^{10}$	$[Xe]4f^{14}5d^9 6s^1$			
Oxidation states	3+, 2+	4+, 2+	4+, 2+			
Electron affinity	1.16	0.562	2.13		eV	Reaction type $Ni + e^- = Ni^-$
Electronegativity χ_A	1.75	1.35	(1.44)			Allred and Rochow
1st ionization energy	7.6398	8.3369	9.0		eV	
2nd ionization energy	18.16884	19.43	18.563		eV	
3rd ionization energy	35.19	32.93			eV	
4th ionization energy	54.9				eV	
Standard electrode potential E^0	−0.257	+0.951	+1.118		V	Reaction type $Ni^{2+} + 2e^- = Ni$

Table 2.1-24B(a) Elements of Group VIII(3) (CAS notation) or Group 10 (new IUPAC notation). Crystallographic properties

Element name	Nickel	Palladium	Platinum			
Chemical symbol Atomic number Z	Ni 28	Pd 46	Pt 78	Uun 110		
					Units	Remarks
Crystal system, Bravais lattice	cub, fc	cub, fc	cub, fc			
Structure type	Cu	Cu	Cu			
Lattice constant a	352.41	389.01	392.33		pm	
Space group	$Fm\bar{3}m$	$Fm\bar{3}m$	$Fm\bar{3}m$			
Schoenflies symbol	O_h^5	O_h^5	O_h^5			
Strukturbericht type	A1	A1	A1			
Pearson symbol	cF4	cF4	cF4			
Number A of atoms per cell	4	4	4			
Coordination number	12	12	12			
Shortest interatomic distance, solid	249	274	277		pm	

Table 2.1-24B(b) Elements of Group VIII(3) (CAS notation), or Group 10 (new IUPAC notation). Mechanical properties

Element name	**Nickel**	**Palladium**	**Platinum**			
Chemical symbol	**Ni**	**Pd**	**Pt**	**Uun**		
Atomic number Z	**28**	**46**	**78**	**110**		
					Units	**Remarks**
Density ϱ, solid	8.90	12.00	21.40		g/cm^3	
Density ϱ, liquid	7.910	10.700	19.700		g/cm^3	
Molar volume V_{mol}	6.59	8.85	9.10		cm^3/mol	
Viscosity η, liquid	5.0				mPa s	
Surface tension, liquid	1.725	1.50	1.866		N/m	
Temperature coefficient	-0.98×10^{-3}	-0.22×10^{-3}	-0.17×10^{-3}		N/(m K)	
Coefficient of linear thermal expansion α	13.3×10^{-6}	11.2×10^{-6}	9.0×10^{-6}		1/K	
Sound velocity, solid, transverse	3080	1900	1690		m/s	
Sound velocity, solid, longitudinal	5810	4540	4080		m/s	
Compressibility κ	0.513×10^{-5}	0.505×10^{-5}	0.351×10^{-5}		1/MPa	Volume compressibility
Elastic modulus E	220	121	170		GPa	
Shear modulus G	78.5	43.5	60.9		GPa	
Poisson number μ	0.31	0.39	0.39			
Elastic compliance s_{11}	7.67	13.7	7.35		1/TPa	
Elastic compliance s_{44}	8.23	14.1	13.1		1/TPa	
Elastic compliance s_{12}	−2.93	−6.0	−3.08		1/TPa	
Elastic stiffness c_{11}	247	221	347		GPa	
Elastic stiffness c_{44}	122	70.8	76.5		GPa	
Elastic stiffness c_{12}	153	171	251		GPa	
Tensile strength	317		134		MPa	At 293 K
Vickers hardness	640	461	560			At 293 K

Table 2.1-24B(c) Elements of Group VIII(3) (CAS notation), or Group 10 (new IUPAC notation). Thermal and thermodynamic properties

Element name	**Nickel**	**Palladium**	**Platinum**			
Chemical symbol	**Ni**	**Pd**	**Pt**	**Uun**		
Atomic number Z	**28**	**46**	**78**	**110**		
					Units	**Remarks**
Thermal conductivity λ	83	71.8	71.6		W/(m K)	
Molar heat capacity c_p	26.07	25.98	25.85		J/(mol K)	At 298 K
Standard entropy S^0	29.796	37.823	41.631		J/(mol K)	At 298 K and 100 kPa
Enthalpy difference $H_{298} - H_0$	4.7870	5.4685	5.7237		kJ/mol	
Melting temperature T_m	1728.30	1828.0	2041.50		K	
Transition	α–liquid	α–liquid	α–liquid			
Enthalpy change ΔH_m	17.4798	16.736	22.1750		kJ/mol	
Entropy change ΔS_m	10.114	9.155	10.862		J/(mol K)	
Relative volume change ΔV_m						$(V_l - V_s)/V_l$ at T_m
Boiling temperature T_b	3157	3237	4100		K	
Enthalpy change ΔH_b	369.24	357.6	509.8		kJ/mol	

Table 2.1–24B(d) Elements of Group VIII(3) (CAS notation), or Group 10 (new IUPAC notation). Electronic, electromagnetic, and optical properties. There is no Table 2.1-24C, because no allotropic or high-pressure modifications are known

Element name	Nickel	Palladium	Platinum			
Chemical symbol	Ni	Pd	Pt	Uun		
Atomic number Z	28	46	78	110		
					Units	Remarks
Characteristics	Ductile metal	Ductile metal	Ductile metal			
Electrical resistivity ρ_s	59.0	101	98.1		nΩ m	At RT
Temperature coefficient	69.2×10^{-4}	37.7×10^{-4}	39.6×10^{-4}		1/K	
Pressure coefficient	1.82×10^{-9}	-2.1×10^{-9}	-1.88×10^{-9}		1/hPa	
Electrical resistivity ρ_l	850				nΩ m	
Resistivity ratio at T_m	1.3					
Hall coefficient R	-60×10^{-12}	-86×10^{-12}	-24.4×10^{-12}		$m^3/(A\,s)$	At 298 K, $B = 0.3–4.6$ T
Thermoelectronic coefficient	−18	−9.54	−3.50		μV/K	
Electronic work function	5.15	5.12	5.65		V	
Thermal work function	4.60	4.99	5.30		V	
Molar magnetic susceptibility χ_{mol} (SI)	Ferromagnetic	6786×10^{-6}	2425×10^{-6}		cm^3/mol	At 295 K
Molar magnetic susceptibility χ_{mol} (cgs)	Ferromagnetic	540×10^{-6}	193×10^{-6}		cm^3/mol	At 295 K
Mass magnetic susceptibility χ_{mass} (SI)	Ferromagnetic	67.0×10^{-6}	13.0×10^{-6}		cm^3/g	At 295 K

Table 2.1–24D Elements of Group VIII(3) (CAS notation), or Group 10 (new IUPAC notation). Ionic radii (determined from crystal structures)

Element	Nickel		Palladium			Platinum		
Ion	Ni^{2+}	Ni^{3+}	Pd^{2+}	Pd^{3+}	Pd^{4+}	Pt^{2+}	Pt^{4+}	
Coordination number								Units
4	49		64			60		pm
6	69	56	86	76	62	80	63	pm

2.1.5.4 Elements of the Lanthanides Period

Table 2.1-25A Lanthanides. Atomic, ionic, and molecular properties (see Table 2.1-25D for ionic radii)

Element name	Cerium	Praseodymium	Neodymium	Promethium	Samarium	Europium	Gadolinium	
Chemical symbol	Ce	Pr	Nd	Pm	Sm	Eu	Gd	
Atomic number Z	58	59	60	61	62	63	64	
								Units
Characteristics				Radioactive				
Relative atomic mass A (atomic weight)	140.116(1)	140.90765(2)	144.24(3)	[145]	150.36(3)	151.964(1)	157.25(3)	
Abundance in lithosphere [a]	41.6×10^{-6}	5.5×10^{-6}	23.9×10^{-6}		6.5×10^{-6}	1.1×10^{-6}	6.4×10^{-6}	
Abundance in sea [a]	1×10^{-12}	6×10^{-13}	3×10^{-12}		5×10^{-14}	1×10^{-14}	7×10^{-13}	
Atomic radius r_{cov}	165	165	164	165	162	185	161	pm
Atomic radius r_{met}	182	183	182	181	180	200	179	pm
Electron shells	–NOP	–NOP	–NOP	–NOP	–NOP	–NOP	–NOP	
Electronic ground state	$^3H_4$	$^4I_{9/2}$	$^5I_4$	$^6H_{5/2}$	$^7F_0$	$^8S_{7/2}$	$^9D_2$	
Electronic configuration	$[Xe]4f^1 5d^1 6s^2$	$[Xe]4f^3 6s^2$	$[Xe]4f^4 6s^2$	$[Xe]4f^5 6s^2$	$[Xe]4f^6 6s^2$	$[Xe]4f^7 6s^2$	$[Xe]4f^7 5d^1 6s^2$	
Oxidation states	4+, 3+	4+, 3+	3+	3+	3+, 2+	3+, 2+	3+	
Electronegativity χ_A [b]	(1.08)	(1.07)	(1.07)	(1.07)	(1.07)	(1.01)	1.11	
1st ionization energy	5.5387	5.464	5.5250	5.55	5.6437	5.6704	6.1500	eV
2nd ionization energy	10.85	10.55	10.73	10.90	11.07	11.241	12.09	eV
3rd ionization energy	20.198	21.624	22.1	22.3	23.4	24.92	20.63	eV
4th ionization energy	36.758	38.98	40.4	41.1	41.4	42.7	44.0	eV

[a] Mass ratio. [b] According to Allred and Rochow.

Element name	Terbium	Dysprosium	Holmium	Erbium	Thulium	Ytterbium	Lutetium	
Chemical symbol	Tb	Dy	Ho	Er	Tm	Yb	Lu	
Atomic number Z	65	66	67	68	69	70	71	
								Units
Relative atomic mass A (atomic weight)	158.92534(2)	162.50(3)	164.93032(2)	167.26(3)	168.93421(2)	173.04(3)	174.967(1)	
Abundance in lithosphere [a]	1×10^{-6}	4.5×10^{-6}	1.2×10^{-6}	2.5×10^{-6}	0.2×10^{-6}	2.7×10^{-6}	0.8×10^{-6}	
Abundance in sea [a]	1×10^{-13}	9×10^{-13}	2×10^{-13}	8×10^{-13}	2×10^{-13}	8×10^{-13}	2×10^{-13}	
Atomic radius r_{cov}	159	159	158	157	156	174	156	pm
Atomic radius r_{met}	176	175	174	173	172	194	172	pm
Electron shells	–NOP	–NOP	–NOP	–NOP	–NOP	–NOP	–NOP	
Electronic ground state	$^6H_{15/2}$	$^5I_8$	$^4I_{15/2}$	$^3H_6$	$^2F_{7/2}$	$^1S_0$	$^2D_{3/2}$	
Electronic configuration	$[Xe]4f^9 6s^2$	$[Xe]4f^{10} 6s^2$	$[Xe]4f^{11} 6s^2$	$[Xe]4f^{12} 6s^2$	$[Xe]4f^{13} 6s^2$	$[Xe]4f^{14} 6s^2$	$[Xe]4f^{14} 5d^1 6s^2$	
Oxidation states	4+, 3+	3+	3+	3+	3+, 2+	3+, 2+	3+	
Electronegativity χ_A [b]	(1.10)	(1.10)	(1.10)	(1.11)	(1.11)	(1.06)	(1.14)	
1st ionization energy	5.8639	5.9389	6.0216	6.1078	6.18431	6.25416	5.42585	eV
2nd ionization energy	11.52	11.67	11.80	11.93	12.05	12.1761	13.9	eV
3rd ionization energy	21.91	22.8	22.84	22.74	23.68	25.05	20.9594	eV
4th ionization energy	39.79	41.4	42.5	42.7	42.7	43.56	45.25	eV

[a] Mass ratio. [b] According to Allred and Rochow.

Table 2.1-25B(a) Lanthanides. Crystallographic properties (see Table 2.1-25C for allotropic and high-pressure modifications)

Element name	Cerium	Praseodymium	Neodymium	Promethium	Samarium	Europium	Gadolinium	
Chemical symbol	Ce	Pr	Nd	Pm	Sm	Eu	Gd	
Atomic number Z	58	59	60	61	62	63	64	
								Units
Modification					α-Sm			
Crystal system, Bravais lattice	cub, fc	hex	hex	hex	hex	cub, bc	hex	
Structure type	Cu	α-La	α-La	α-La	Se	W	Mg	
Lattice constant a	516.10	367.21	365.82	365	362.90	458.27	363.36	pm
Lattice constant c		1183.26	1179.66	1165	2620.7		578.10	pm
Space group	$Fm\bar{3}m$	$P6_3/mmc$	$P6_3/mmc$	$P6_3/mmc$		$Im\bar{3}m$	$P6_3/mmc$	
Schoenflies symbol	O_h^5	D_{6h}^4	D_{6h}^4	D_{6h}^4		O_h^9	D_{6h}^4	
Strukturbericht type	A1	A3′	A3′	A3′		A2	A3	
Pearson symbol	cF4	hP4	hP4	hP4		cI2	hP2	
Number A of atoms per cell	4	4	4	4		2	2	
Coordination number	12	12	12	12			12	
Shortest interatomic distance, solid	364		364	362		397	360	pm

Element name	Terbium	Dysprosium	Holmium	Erbium	Thulium	Ytterbium	Lutetium	
Chemical symbol	Tb	Dy	Ho	Er	Tm	Yb	Lu	
Atomic number Z	65	66	67	68	69	70	71	
								Units
Crystal system, Bravais lattice	hex	hex	hex	hex	hex	cub, fc	hex	
Structure type	Mg	Mg	Mg	Mg	Mg	Cu	Mg	
Lattice constant a	360.55	359.15	357.78	355.92	353.75	548.48	350.52	pm
Lattice constant c	569.66	565.01	561.78	558.50	555.40		554.94	pm
Space group	$P6_3/mmc$	$P6_3/mmc$	$P6_3/mmc$	$P6_3/mmc$	$P6_3/mmc$	$Fm\bar{3}m$	$P6_3/mmc$	
Schoenflies symbol	D_{6h}^4	D_{6h}^4	D_{6h}^4	D_{6h}^4	D_{6h}^4	O_h^5	D_{6h}^4	
Strukturbericht type	A3	A3	A3	A3	A3	A1	A3	
Pearson symbol	hP2	hP2	hP2	hP2	hP2	cF4	hP2	
Number A of atoms per cell	2	2	2	2	2	4	2	
Coordination number	12	12	12	12	12	12	12	
Shortest interatomic distance, solid	357	355	353	351	349	388	347	pm

Table 2.1-25B(b) Lanthanides. Mechanical properties

Element name	Cerium [a]	Praseodymium	Neodymium	Promethium	Samarium	Europium	Gadolinium	
Chemical symbol	Ce	Pr	Nd	Pm	Sm	Eu	Gd	
Atomic number Z	58	59	60	61	62	63	64	
								Units
Density ϱ, solid	6.78	6.77	7.00	6.48	7.54	5.26	7.89	g/cm^3
Density ϱ, liquid		6.609						g/cm^3
Molar volume V_{mol}	17.00	20.80	20.59	20.1	20.00	28.29	19.90	cm^3/mol
Viscosity η, liquid		0.431						mPa s
Surface tension, liquid	0.72	0.7	0.688	0.65	0.6	0.450	0.816	N/m
Coefficient of linear thermal expansion α	6.3×10^{-6}	6.79×10^{-6}	9.6×10^{-6}	11×10^{-6}	10.4×10^{-6}	32×10^{-6}	9.4×10^{-6}	1/K
Sound velocity, solid, transverse	1230	1410	1440		1290		1680	m/s
Sound velocity, solid, longitudinal	3060	2660	2720		2700		2950	m/s
Volume compressibility κ	5.06×10^{-5}	3.15×10^{-5}	2.94×10^{-5}	2.9×10^{-5}	3.27×10^{-5}	8.13×10^{-5}	2.47×10^{-5}	1/MPa
Elastic modulus E	44.1	35.2	37.9	46	53.9	18.2	56.3	GPa
Shear modulus G	12.1	13.2	14.3	18	12.7	7.8	22.4	GPa
Poisson number μ	0.25	0.31	0.31	0.28	0.35	0.17	0.26	
Elastic compliance s_{11}	62.8	26.6	23.7				18.0	1/TPa
Elastic compliance s_{33}		19.3	18.5				16.1	1/TPa
Elastic compliance s_{44}	57.8	73.6	66.5				48.1	1/TPa
Elastic compliance s_{12}	−22.3	−11.3	−9.5				−5.7	1/TPa
Elastic compliance s_{13}		−3.8	−3.9				−3.6	1/TPa
Elastic stiffness c_{11}	26.0	49.4	54.8				67.8	GPa
Elastic stiffness c_{33}		57.4	60.9				71.2	GPa
Elastic stiffness c_{44}	17.3	13.6	15.0				20.8	GPa
Elastic stiffness c_{12}	14.3	23.0	24.6				25.6	GPa
Elastic stiffness c_{13}		14.3	16.6				20.7	GPa
Tensile strength	117	169	169	170	157		122	MPa
Vickers hardness	275	400	348		412	167	510–638	

[a] The data for the elastic behavior of cerium concern γ-cerium.

Table 2.1-25B(b) Lanthanides. Mechanical properties, cont.

Element name	Terbium	Dysprosium	Holmium	Erbium	Thulium	Ytterbium	Lutetium	
Chemical symbol	**Tb**	**Dy**	**Ho**	**Er**	**Tm**	**Yb**	**Lu**	
Atomic number Z	**65**	**66**	**67**	**68**	**69**	**70**	**71**	
								Units
Density ϱ, solid	8.27	8.54	8.80	9.05	9.33	6.98	9.840	g/cm^3
Density ϱ, liquid						6.292		g/cm^3
Molar volume V_{mol}	19.31	19.00	18.75	18.44	18.12	24.84	17.78	cm^3/mol
Viscosity η, liquid						0.424		mPa s
Surface tension, liquid	0.65	0.650	0.650	0.620	0.62	0.85		N/m
Coefficient of linear thermal expansion α	7.0×10^{-6}	10.0×10^{-6}	11.2×10^{-6}	9.2×10^{-6}	13.3×10^{-6}	25.0×10^{-6}	8.12×10^{-6}	1/K
Sound velocity, solid, transverse	1060	1720	1740	1810		1000		m/s
Sound velocity, solid, longitudinal	2920	2960	3040	3080		1820		m/s
Volume compressibility κ	2.40×10^{-5}	2.50×10^{-5}	2.42×10^{-5}	2.34×10^{-5}	2.42×10^{-5}	7.24×10^{-5}	2.33×10^{-5}	1/MPa
Elastic modulus E	57.5	63.1	67.1	73.3	74.0	18.4	68.4	GPa
Shear modulus G	22.8	25.5	26.7	29.6	30.4	7.16	27.1	GPa
Poisson number μ	0.26	0.24	0.26	0.24	0.27	0.28	0.27	
Elastic compliance s_{11}	17.4	16.0	15.3	14.1		89.2	14.3	1/TPa
Elastic compliance s_{33}	15.6	14.5	14.0	13.2			14.8	1/TPa
Elastic compliance s_{44}	46.0	41.2	38.6	36.4		56.4	37.3	1/TPa
Elastic compliance s_{12}	−5.2	−4.6	−4.3	−4.2		−31.9	−4.2	1/TPa
Elastic compliance s_{13}	−3.6	−3.2	−2.9	−2.6			−3.5	1/TPa
Elastic stiffness c_{11}	69.2	74.0	76.5	84.1		18.6	86.2	GPa
Elastic stiffness c_{33}	74.4	78.6	79.6	84.7			80.9	GPa
Elastic stiffness c_{44}	21.8	24.3	25.9	27.4		17.7	26.8	GPa
Elastic stiffness c_{12}	25.0	25.5	25.6	29.4		10.4	32.0	GPa
Elastic stiffness c_{13}	21.8	21.8	21.0	22.6			28.0	GPa
Tensile strength	About 122		About 132	139	About 140	72.5	139	MPa
Vickers hardness	863	544	481	589	520	206	1160	

Table 2.1–25B(c) Lanthanides. Thermal and thermodynamic properties

Element name	Cerium	Praseodymium	Neodymium	Promethium	Samarium	Europium	Gadolinium		
Chemical symbol	Ce	Pr	Nd	Pm	Sm	Eu	Gd		
Atomic number Z	58	59	60	61	62	63	64		
								Units	Remarks
Thermal conductivity λ	11.4	12.5	16.5	17.9	13.3	13.9	10.6	W/(m K)	
Molar heat capacity c_p	26.4	27.20	27.45		29.54	27.6	37.02	J/(mol K)	At 298.2 K
Standard entropy S^0	69.454	73.931	71.086	72.00	69.496	80.793	68.089	J/(mol K)	At 298.15 K and 100 kPa
Enthalpy difference $H_{298} - H_0$	7.2802	7.4182	7.1337	7.3000	7.5730	8.0040	9.0876	kJ/mol	At 298.15 K
Melting temperature T_m	1072.0	1204.0	1289.0	1315.0	1345.0	1095.0	1586.15	K	
Transition	δ–liquid	β–liquid	β–liquid	β–liquid	γ–liquid	α–liquid	β–liquid		
Enthalpy change ΔH_m	5.4601	6.8869	7.1421	7.7000	8.6190	9.2132	9.6680	kJ/mol	
Entropy change ΔS_m	5.093	5.720	5.541	5.856	6.408	8.414	6.095	J/(mol K)	
Relative volume change ΔV_m	0.011	0.02	0.09		0.036	0.048	0.02		$(V_l - V_s)/V_l$ at T_m
Boiling temperature T_b	3699	3785	3341	3785	2064	1870	3569	K	
Enthalpy change ΔH_b	414.2	296.8	273.0		166.4	144.7	359.4	kJ/mol	

Table 2.1–25B(c) Lanthanides. Thermal and thermodynamic properties, cont.

Element name	Terbium	Dysprosium	Holmium	Erbium	Thulium	Ytterbium	Lutetium		
Chemical symbol	Tb	Dy	Ho	Er	Tm	Yb	Lu		
Atomic number Z	65	66	67	68	69	70	71		
								Units	Remarks
Thermal conductivity λ	11.1	10.7	16.2	14	16.8	38.5	16.4	W/(m K)	
Molar heat capacity c_p	28.91	28.16	27.15	28.12	27.03	26.74	26.86	J/(mol K)	At 298.2 K
Standard entropy S^0	73.8	74.956	75.019	73.178	74.015	59.831	50.961	J/(mol K)	At 298.15 K and 100 kPa
Enthalpy difference $H_{298} - H_0$	9.4266	8.8659	7.9956	7.3923	7.3973	6.7111	6.3890	kJ/mol	At 298.15 K
Melting temperature T_m	1632	1685.15	1745.0	1802.0	1818.0	1097	1936.0	K	
Transition	β–liquid	β–liquid	β–liquid	α–liquid	α–liquid	γ–liquid	α–liquid		
Enthalpy change ΔH_m	10.1504	11.3505	11.7570	19.9033	16.8406	7.6567	18.6481	kJ/mol	
Entropy change ΔS_m	6.220	6.736	6.738	11.045	9.263	6.980	9.632	J/(mol K)	
Relative volume change ΔV_m	0.031	0.45	0.074	0.09	0.069	0.051	0.036		$(V_l - V_s)/V_l$ at T_m
Boiling temperature T_b	3496	2835	2968	3136	2220	1467	3668	K	
Enthalpy change ΔH_b	330.9	230	242.50	261.4	190.7	128.83	355.9	kJ/mol	

Table 2.1-25B(d) Lanthanides. Electronic, electromagnetic, and optical properties

Element name	Cerium	Praseo-dymium	Neodymium	Prome-thium	Samarium	Europium	Gadolinium		
Chemical symbol	Ce	Pr	Nd	Pm	Sm	Eu	Gd		
Atomic number Z	58	59	60	61	62	63	64		
								Units	Remarks
Characteristics	Metal	Ductile reactive metal	Reactive metal	Metal	Metal	Soft metal very reactive	Soft metal		
Electrical resistivity ρ_s	730	650	610	500	914	890	1260	nΩ m	At RT
Temperature coefficient	9.7×10^{-4}	17.1×10^{-4}	21.3×10^{-4}	28×10^{-4}	14.8×10^{-4}	81.3×10^{-4}	17.6×10^{-4}	1/K	
Pressure coefficient	-45.2×10^{-9}	-0.4×10^{-9}	-1.5×10^{-9}		-3.57×10^{-9}		-4.5×10^{-9}	1/hPa	
Electrical resistivity ρ_l		1130	1550			2440	1950	nΩ m	
Hall coefficient R	1.92×10^{-10}	0.709×10^{-10}	0.971×10^{-10}		-0.2×10^{-10}		0.95×10^{-10}	$m^3/(A\,s)$	At 293 K, $B < 1$ T
Thermoelectric coefficient	4.39							μV/K	
Electronic work function	2.88		3.2		2.7	2.5	3.1	V	
Thermal work function	2.6	2.7	3.3		3.2	2.54	3.07	V	
Molar magnetic susceptibility χ_{mol} (SI)	31.4×10^{-3}	69.5×10^{-3}	74.5×10^{-3}		16.1×10^{-3}	388×10^{-3}	2.324	cm^3/mol	At 295 K
Molar magnetic susceptibility χ_{mol} (cgs)	2.5×10^{-3}	5.53×10^{-3}	5.93×10^{-3}		1.28×10^{-3}	30.9×10^{-3}	185×10^{-3}	cm^3/mol	At 295 K
Mass magnetic susceptibility χ_{mass} (SI)	217×10^{-6}	474×10^{-6}	490×10^{-6}		106×10^{-6}	2.81×10^{-3}	14.9×10^{-3}	cm^3/g	At 295 K

Element name	Terbium	Dysprosium	Holmium	Erbium	Thulium	Ytterbium	Lutetium		
Chemical symbol	Tb	Dy	Ho	Er	Tm	Yb	Lu		
Atomic number Z	65	66	67	68	69	70	71		
								Units	Remarks
Characteristics	Ductile metal	Metal	Soft metal	Soft metal	Ductile metal	Ductile metal	Metal		
Electrical resistivity ρ_s	1130	890	814	810	670	250	540	nΩ m	At RT
Temperature coefficient	11.9×10^{-4}	11.9×10^{-4}	17.1×10^{-4}	20.1×10^{-4}	19.5×10^{-4}	13×10^{-4}	24×10^{-4}	1/K	
Pressure coefficient		-2.3×10^{-9}	-2.2×10^{-9}	-27×10^{-9}	-2.6×10^{-9}	9.7×10^{-9}	-1.31×10^{-9}	1/hPa	
Electrical resistivity ρ_l	1930	2100	2210	2260		1080		nΩ m	
Hall coefficient R				-0.34×10^{-10}		-0.53×10^{-10}		$m^3/(A\,s)$	At 293 K, $B < 1$ T
Electronic work function							3.3	V	
Thermal work function	3.4	3.09	3.09	3.12	3.12	2.50	3.14	V	
Molar magnetic susceptibility χ_{mol} (SI)	2.136 [a]	1.232 [a]	916×10^{-3}	603×10^{-3}	329×10^{-3}	842×10^{-6}	2.30×10^{-3}	cm^3/mol	At 295 K
Molar magnetic susceptibility χ_{mol} (cgs)	170×10^{-3} [a]	98×10^{-3} [a]	72.9×10^{-3}	48×10^{-3}	26.2×10^{-3}	67×10^{-6}	183×10^{-6}	cm^3/mol	At 295 K
Mass magnetic susceptibility χ_{mass} (SI)	15.3×10^{-3}	8.00×10^{-3}	5.49×10^{-3}	3.33×10^{-3}	1.90×10^{-3}	4.9×10^{-6}	13×10^{-6}	cm^3/g	At 295 K

[a] The molar magnetic susceptibility is given for Tb and Dy for the α-phase [1.3].

Table 2.1–25C Lanthanides. Allotropic and high-pressure modifications

Element	Cerium					Praseodymium			
Modification	α-Ce	β-Ce	γ-Ce	α′-Ce	Ce-III	α-Pr	β-Pr	γ-Pr	
									Units
Crystal system, Bravais lattice	cub, fc	hex	cub, fc	cub, fc	orth	hex	cub, bc	cub, fc	
Structure type	Cu	α-La	Cu	Cu	α-U	α-La	W	Cu	
Lattice constant *a*	516.10	367.3		482		367.21	413	488	pm
Lattice constant *c*		1180.2				1183.26			pm
Space group	$Fm\bar{3}m$	$P6_3/mmc$	$Fm\bar{3}m$	$Fm\bar{3}m$	$Cmcm$	$P6_3/mmc$	$Im\bar{3}m$	$Fm\bar{3}m$	
Schoenflies symbol	O_h^5	D_{6h}^4	O_h^5	O_h^5	D_{2h}^{17}	D_{6h}^4	O_h^9	O_h^5	
Strukturbericht type	A1	A3′	A1	A1	A20	A3′	A2	A1	
Pearson symbol	cF4	hP4	cF4	cF4	oC4	hP4	cI2	cF4	
Number *A* of atoms per cell	4	4	4	4	4	4	2	4	
Coordination number	12	12	12	12	2+2 +4+4	12	8	12	
Shortest interatomic distance, solid	364						358	345	pm
Range of stability	RTP	>263 K	<95 K	>1.5 GPa	5.1 GPa	RTP	>1094 K	>4.0 GPa	

Element	Neodymium			Promethium		
Modification	α-Nd	β-Nd	γ-Nd	α-Pm	β-Pm	
						Units
Crystal system, Bravais lattice	hex	cub, bc	cub, fc	hex	cub, bc	
Structure type	α-La	W	Cu	α-La	W	
Lattice constant *a*	365.82	413	480	365		pm
Lattice constant *c*	1179.66			1165		pm
Space group	$P6_3/mmc$	$Im\bar{3}m$	$Fm\bar{3}m$	$P6_3/mmc$	$Im\bar{3}m$	
Schoenflies symbol	D_{6h}^4	O_h^9	O_h^5	D_{6h}^4	O_h^9	
Strukturbericht type	A3′	A2	A1	A3′	A2	
Pearson symbol	hP4	cI2	cF4	hP4	cI2	
Number *A* of atoms per cell	4	2	4	4	2	
Coordination number	12	8	12	12	8	
Shortest interatomic distance, solid	364	358	339	362		pm
Range of stability	RTP	> 1135 K	> 5.0 GPa	RTP	> 1163 K	

Element	Samarium			Gadolinium			
Modification	α-Sm	β-Sm	γ-Sm	α-Gd	β-Gd	γ-Gd	
							Units
Crystal system, Bravais lattice	trig	cub, bc	hex	hex, cp	cub, bc	trig	
Structure type	α-Sm	W	α-La	Mg	W	α-Sm	
Lattice constant *a*	362.90		361.8	363.36	406	361	pm
Lattice constant *c*	2620.7		1166	578.10		2603	pm
Space group	$R\bar{3}m$	$Im\bar{3}m$	$P6_3/mmc$	$P6_3/mmc$	$Im\bar{3}m$	$R\bar{3}m$	
Schoenflies symbol	D_{3d}^5	O_h^9	D_{6h}^4	D_{6h}^4	O_h^9	D_{3d}^5	
Strukturbericht type	A7	A2	A3′	A3	A2	A7	
Pearson symbol	hR3	cI2	hP4	hP2	cI2	hR3	
Number *A* of atoms per cell	3	2	4	2	2	3	
Coordination number	12	8	12	12	8		
Shortest interatomic distance, solid	361	353	360	360	351		pm
Range of stability	RTP	> 1190 K	> 4.0 GPa	RTP	> 1535 K	> 3.0 GPa	

Table 2.1-25C Lanthanides. Allotropic and high-pressure modifications, cont.

Element	Terbium			Dysprosium				
Modification	α-Tb	β-Tb	Tb-II	α-Dy	β-Dy	α′-Dy	γ-Dy	
								Units
Crystal system, Bravais lattice	hex, cp	cub, bc	trig	hex, cp	cub, bc	orth	trig	
Structure type	Mg	W	α-Sm	Mg	W		α-Sm	
Lattice constant a	360.55		341	359.15		359.5	343.6	pm
Lattice constant b						618.4		pm
Lattice constant c	569.66		2450	565.01		567.8	2483.0	pm
Space group	$P6_3/mmc$	$Im\bar{3}m$	$R\bar{3}m$	$P6_3/mmc$	$Im\bar{3}m$	$Cmcm$	$R\bar{3}m$	
Schoenflies symbol	D_{6h}^4	O_h^9	D_{3d}^5	D_{6h}^4	O_h^9	D_{2h}^{17}	D_{3d}^5	
Strukturbericht type	A3	A2	A7	A3	A2			
Pearson symbol	hP2	cI2	hR3	hP2	cI2	oC4	hR3	
Number A of atoms per cell	2	2	3	2	2	4	3	
Coordination number	12			12	8			
Shortest interatomic distance, solid	357			355	345			pm
Range of stability	RTP	> 1589 K	> 6.0 GPa	RTP	> 1243 K	< 86 K	> 7.5 GPa	

Element	Holmium			Erbium		Thulium			
Modification	α-Ho	β-Ho	γ-Ho	α-Er	β-Er	α-Tm	β-Tm	Tm-II	
									Units
Crystal system, Bravais lattice	hex, cp	cub, bc	trig	hex, cp	cub, bc	hex, cp	cub, bc	trig	
Structure type	Mg	W	α-Sm	Mg	W	Mg	W	α-Sm	
Lattice constant a	357.78		334	355.92		353.75			pm
Lattice constant c	561.78		2450	558.50		555.40			pm
Space group	$P6_3/mmc$	$Im\bar{3}m$	$R\bar{3}m$	$P6_3/mmc$	$Im\bar{3}m$	$P6_3/mmc$	$Im\bar{3}m$	$R\bar{3}m$	
Schoenflies symbol	D_{6h}^4	O_h^9	D_{3d}^5	D_{6h}^4	O_h^9	D_{6h}^4	O_h^9	D_{3d}^5	
Strukturbericht type	A3	A2	A7	A3	A2	A3	A2		
Pearson symbol	hP2	cI2	hR3	hP2	cI2	hP2	cI2	hR3	
Number A of atoms per cell	2	2	3	2	2	2	2	3	
Coordination number	12	8		12	8	12	8		
Shortest interatomic distance, solid	353	343		351		349			pm
Range of stability	RTP	High temperature	> 4.0 GPa	RTP	High temperature	RTP	High temperature	> 6.0 GPa	

Element	Ytterbium			Lutetium			
Modification	α-Yb	β-Yb	γ-Yb	α-Lu	β-Lu	Lu-II	
							Units
Crystal system, Bravais lattice	cub, fc	cub, bc	hex, cp	hex, cp	cub, bc	trig	
Structure type	Cu	W	Mg	Mg	W	α-Sm	
Lattice constant a	548.48	444	387.99	350.52			pm
Lattice constant c			638.59	554.94			pm
Space group	$Fm\bar{3}m$	$Im\bar{3}m$	$P6_3/mmc$	$P6_3/mmc$	$Im\bar{3}m$	$R\bar{3}m$	
Schoenflies symbol	O_h^5	O_h^9	D_{6h}^4	D_{6h}^4	O_h^9	D_{3d}^5	
Strukturbericht type	A1	A2	A3	A3	A2		
Pearson symbol	cF4	cI2	hP2	hP2	cI2	hR3	
Number A of atoms per cell	4	2	2	2	2	3	
Coordination number	12	8	6+6	12	8	12	
Shortest interatomic distance, solid	388	385	388	347	385		pm
Range of stability	RTP	> 1005 K	< 270 K	RTP	> 1005 K	> 23 GPa	

Table 2.1–25D Lanthanides. Ionic radii (determined from crystal structures)

Element	Cerium		Praseo-dymium		Neo-dymium	Prome-thium	Samarium		Europium		Gado-linium	
Ion	Ce^{3+}	Ce^{4+}	Pr^{3+}	Pr^{4+}	Nd^{3+}	Pm^{3+}	Sm^{2+}	Sm^{3+}	Eu^{2+}	Eu^{3+}	Gd^{3+}	
Coordination number												**Units**
6	101	87	99	85	98		119	96	117	95	94	pm
8	114	97	113	96	112	109	127	108	125	107	105	pm
9					116							pm
10	125	107							135			pm
12	134	114			127			124				pm

Element	Terbium		Dysprosium		Erb-ium	Thulium		Ytterbium		Lute-tium	
Ion	Tb^{3+}	Tb^{4+}	Dy^{2+}	Dy^{3+}	Er^{3+}	Tm^{2+}	Tm^{3+}	Yb^{2+}	Yb^{3+}	Lu^{3+}	
Coordination number											**Units**
6	92	76	107	91	89	101	88	102		86	pm
7						109					pm
8	104	88	119	103	100		99	114	99	97	pm
9									104		pm

2.1.5.5 Elements of the Actinides Period

Table 2.1-26A Actinides. Atomic, ionic, and molecular properties (see Table 2.1-26D for ionic radii)

Element name	Thorium	Protactinium	Uranium	Neptunium	Plutonium	Americium	Curium	
Chemical symbol	**Th**	**Pa**	**U**	**Np**	**Pu**	**Am**	**Cu**	
Atomic number Z	**90**	**91**	**92**	**93**	**94**	**95**	**96**	
								Units
Characteristics	Radioactive	Radioactive	Radioactive	Radioactive	Radioactive	Radioactive	Radioactive	
Relative atomic mass A (atomic weight)	232.0381(1)	231.03588(2)	238.0289(1)	[237]	[244]	[243]	[247]	
Abundance in lithosphere [a]	11.5×10^{-6}		4×10^{-6}					
Atomic radius r_{cov}	165		142					pm
Atomic radius r_{met}	180	164	154	150	164	173	174	pm
Electron shells	–OPQ	–OPQ	–OPQ	–OPQ	–OPQ	–OPQ	–OPQ	
Electronic ground state	$^3F_2$	$^4K_{11/2}$	$^5L_6$	$^6L_{11/2}$	$^7F_0$	$^8S_{7/2}$	$^9D_2$	
Electronic configuration	$[Rn]6d^27s^2$	$[Rn]5f^26d^17s^2$	$[Rn]5f^36d^17s^2$	$[Rn]5f^46d^17s^2$	$[Rn]5f^67s^2$	$[Rn]5f^77s^2$	$[Rn]5f^76d^17s^2$	
Oxidation states	4+	5+, 4+	6+, 5+, 4+, 3+	6+, 5+, 4+, 3+	6+, 5+, 4+, 3+	6+, 5+, 4+, 3+	3+	
Electronegativity χ_A [b]	(1.11)	(1.14)	(1.22)	(1.22)	(1.22)	(1.2)	(1.2)	
1st ionization energy	6.08	5.89	6.19405	6.2657	6.06	5.993	6.02	eV
2nd ionization energy	11.5							eV
3rd ionization energy	20.0							eV
4th ionization energy	28.8							eV
Standard electrode potential E^0 [c]	−1.899							V

[a] Mass ratio. [b] According to Allred and Rochow. [c] Reaction type $Th^{4+} + 4e^- = Th$

Element name	Berkelium	Californium	Einsteinium	Fermium	Mendelevium	Nobelium	Lawrencium	
Chemical symbol	**Bk**	**Cf**	**Es**	**Fm**	**Md**	**No**	**Lr**	
Atomic number Z	**97**	**98**	**99**	**100**	**101**	**102**	**103**	
								Units
Characteristics	Radioactive	Radioactive	Radioactive	Radioactive	Radioactive	Radioactive	Radioactive	
Relative atomic mass A (atomic weight)	[247]	[251]	[252]	[257]	[258]	[259]	[262]	
Atomic radius r_{met}	170	186	186					pm
Electron shells	–OPQ	–OPQ	–OPQ	–OPQ	–OPQ	–OPQ	–OPQ	
Electronic ground state	$^6H_{15/2}$	$^5I_8$	$^5I_{15/2}$	$^3H_6$	$^2F_{7/2}$	$^1S_0$	$^2D_{5/2}$	
Electronic configuration	$[Rn]5f^97s^2$	$[Rn]5f^{10}7s^2$	$[Rn]5f^{11}7s^2$	$[Rn]5f^{12}7s^2$	$[Rn]5f^{13}7s^2$	$[Rn]5f^{14}7s^2$	$[Rn]5f^{14}6d^17s^2$	
Oxidation states	4+, 3+	3+					3+	
Electronegativity χ_A [a]	(1.2)	(1.2)	(1.2)	(1.2)	(1.2)			
1st ionization energy	6.23	6.30	6.42	6.50	6.58	6.65		eV

[a] According to Allred and Rochow.

Table 2.1-26B(a) Actinides. Crystallographic properties (see Table 2.1-26C for allotropic and high-pressure modifications)

Element name	Thorium	Protactinium	Uranium	Neptunium	Plutonium	Americium	Curium	
Chemical symbol	**Th**	**Pa**	**U**	**Np**	**Pu**	**Am**	**Cu**	
Atomic number Z	**90**	**91**	**92**	**93**	**94**	**95**	**96**	
								Units
Modification			α-U		α-Pu			
Crystal system, Bravais lattice	cub, fc	tetr, I	orth, C	orth, P	mon, P	hex	hex	
Structure type	Cu	α-Pa	α-U	α-Np	α-Pu	α-La	α-La	
Lattice constant *a*	508.51	394.5	285.38	666.3	618.3	346.8	349.6	pm
Lattice constant *b*			586.80	472.3	482.2			pm
Lattice constant *c*		324.2	495.57	488.7	1096.8	1124.1	1133.1	pm
Lattice angle γ					101.78			deg
Space group	$Fm3m$	$I4/mmm$	$Cmcm$	$Pnma$	$P2_1/m$	$P6_3/mmc$	$P6_3/mmc$	
Schoenflies symbol	O_h^5	D_{4h}^{17}	D_{2h}^{17}	D_{2h}^{16}	C_{2h}^2	D_{6h}^4	D_{6h}^4	
Strukturbericht type	A1	A6	A20	A_c		A3′	A3′	
Pearson symbol	cF4	tI2	oC4	oP8	mP16	hP4	hP4	
Number *A* of atoms per cell	4	2	4	8	16	4	4	
Coordination number	12	8+2	2+2+4+4	8+6		6+6	6+6	
Shortest interatomic distance, solid	360	321	276	259–335	257–278	347	349	pm

Element name	Berkelium	Californium	Einsteinium	Fermium	Mendelevium	Nobelium	Lawrencium	
Chemical symbol	**Bk**	**Cf**	**Es**	**Fm**	**Md**	**No**	**Lr**	
Atomic number Z	**97**	**98**	**99**	**100**	**101**	**102**	**103**	
								Units
Crystal system, Bravais lattice	hex	hex	hex					
Structure type	α-La	α-La	α-La					
Lattice constant *a*	341.6							pm
Lattice constant *c*	1106.9							pm
Space group	$P6_3/mmc$	$P6_3/mmc$	$P6_3/mmc$					
Schoenflies symbol	D_{6h}^4	D_{6h}^4	D_{6h}^4					
Strukturbericht type	A3′	A3′	A3′					
Pearson symbol	hP4	hP4	hP4					
Number *A* of atoms per cell	4	4	4					

Table 2.1–26B(b) Actinides. Mechanical properties

Element name	Thorium	Protactinium	Uranium	Neptunium	Plutonium	Americium	Curium	
Chemical symbol	**Th**	**Pa**	**U**	**Np**	**Pu**	**Am**	**Cu**	
Atomic number Z	**90**	**91**	**92**	**93**	**94**	**95**	**96**	
								Units
Modification					α-Pu			
Density ϱ, solid	11.70	15.40	18.90	20.40	19.80	13.60	13.510	g/cm^3
Density ϱ, liquid	10.35		17.907		16.623			
Molar volume V_{mol}	19.80	15.0	12.56	11.71	12.3	17.78	18.3	cm^3/mol
Viscosity η, liquid					7.4			mPa s
Surface tension, liquid	1.05		1.53		0.55			N/m
Temperature coefficient			-0.14×10^{-3}					N/(m K)
Coefficient of linear thermal expansion α	12.5×10^{-6}	7.3×10^{-6}	12.6×10^{-6}	27.5×10^{-6}	55×10^{-6}			1/K
Sound velocity, solid, transverse	1630		1940					m/s
Sound velocity, solid, longitudinal	2850		3370					m/s
Volume compressibility κ	1.86×10^{-5}		0.785×10^{-5}					1/MPa
Elastic modulus E	78.3	76	177	68	92.7			GPa
Shear modulus G	30.8		70.6		45			GPa
Poisson number μ	0.26		0.25		0.18			
Elastic compliance s_{11}	27.4		4.91					1/TPa
Elastic compliance s_{22}			6.73					1/TPa
Elastic compliance s_{33}			4.79					1/TPa
Elastic compliance s_{44}	22.0		8.04					1/TPa
Elastic compliance s_{55}			13.6					1/TPa
Elastic compliance s_{66}			13.4					1/TPa
Elastic compliance s_{12}	−10.9		−1.19					1/TPa
Elastic compliance s_{13}			0.08					1/TPa
Elastic compliance s_{23}			−2.61					1/TPa
Elastic stiffness c_{11}	77.0		215					GPa
Elastic stiffness c_{22}			199					GPa
Elastic stiffness c_{33}			267					GPa
Elastic stiffness c_{44}	45.5		124					GPa
Elastic stiffness c_{55}			73.4					GPa
Elastic stiffness c_{66}			74.3					GPa
Elastic stiffness c_{12}	50.9		46					GPa
Elastic stiffness c_{13}			22					GPa
Elastic stiffness c_{23}			108					GPa
Tensile strength	0.219		0.585		0.525			GPa
Vickers hardness	294–687		1960		2500–2800			

Table 2.1-26B(b) Actinides. Mechanical properties, cont.

Element name	Berkelium	Californium	Einsteinium	Fermium	Mendelevium	Nobelium	Lawrencium	
Chemical symbol	**Bk**	**Cf**	**Es**	**Fm**	**Md**	**No**	**Lr**	
Atomic number Z	**97**	**98**	**99**	**100**	**101**	**102**	**103**	
								Units
Density ϱ, solid	14.790	9.310	8.840					g/cm^3
Molar volume V_{mol}	16.70	26.96	28.5					cm^3/mol

Table 2.1-26B(c) Actinides. Thermal and thermodynamic properties

Element name	Thorium	Protactinium	Uranium	Neptunium	Plutonium	Americium	Curium		
Chemical symbol	**Th**	**Pa**	**U**	**Np**	**Pu**	**Am**	**Cu**		
Atomic number Z	**90**	**91**	**92**	**93**	**94**	**95**	**96**		
								Units	**Remarks**
Modification	α-Th	α-Pa	α-U	α-Np	α-Pu				
Thermal conductivity λ	77.0	About 47	22.0	6.3	6.74	About 10	About 10	W/(m K)	
Molar heat capacity c_p	27.32	(27.61)	27.66	29.62	32.84	25.9	(27.70)	J/(mol K)	At 298 K
Standard entropy S^0	51.800	51.882	50.200	50.459	54.461	55.396	71.965	J/(mol K)	At 298.15 K and 100 kPa
Enthalpy difference $H_{298} - H_0$	6.3500	6.4392	6.3640	6.6065	6.9023	6.4070	6.1337	kJ/mol	
Melting temperature T_m	2022.99	1844.78	1408.0	917.0	913.0	1449.0	1618.0	K	
Transition	β–liquid	β–liquid	γ–liquid	γ–liquid	ε–liquid	γ–liquid	β–liquid		
Enthalpy change ΔH_m	13.8072	12.3412	9.1420	3.1986	2.8240	14.3930	14.6440	kJ/mol	
Entropy change ΔS_m	6.825	6.690	6.493	3.488	3.093	9.933	9.051	J/(mol K)	
Relative volume change ΔV_m	0.05		0.022						$(V_l - V_s)/V_l$ at T_m
Boiling temperature T_b	5061		4407		3503	2880		K	
Enthalpy change ΔH_b	514.1	481	464.1	423.4	260.0	238.5	395.7	kJ/mol	

Element name	Berkelium	Californium	Einsteinium	Fermium	Mendelevium	Nobelium	Lawrencium		
Chemical symbol	**Bk**	**Cf**	**Es**	**Fm**	**Md**	**No**	**Lr**		
Atomic number Z	**97**	**98**	**99**	**100**	**101**	**102**	**103**		
								Units	**Remarks**
Thermal conductivity λ	About 10	About 10	About 10					W/(m K)	
Standard entropy S^0	(76.15)	80.542	89.471	87.236				J/(mol K)	At 298.15 K and 100 kPa
Melting temperature T_m	1256	1213	1133.0	1800	1100	1100	1900	K	
Transition			β–liquid						
Enthalpy change ΔH_m			9.4056	1.02				kJ/mol	
Entropy change ΔS_m			8.302					J/(mol K)	
Enthalpy of boiling ΔH_b				3.26				kJ/mol	

Table 2.1-26B(d) Actinides. Electronic, electromagnetic, and optical properties

Element name	Thorium	Protactinium	Uranium	Neptunium	Plutonium	Americium	Curium		
Chemical symbol	**Th**	**Pa**	**U**	**Np**	**Pu**	**Am**	**Cu**		
Atomic number Z	**90**	**91**	**92**	**93**	**94**	**95**	**96**		
								Units	**Remarks**
Characteristics	Soft metal	Toxic metal	Ductile metal	Reactive metal	Metal, very toxic	Metal	Metal, very reactive		
Electrical resistivity ρ_s	147	177	280		1414	680	860	nΩ m	At RT
Temperature coefficient	27.5×10^{-4}		28.2×10^{-4}			33×10^{-4}		1/K	
Pressure coefficient	-3.4×10^{-9}							1/Pa	
Superconducting critical temperature T_{crit}	1.37							K	
Superconducting critical field H_{crit}	162							Oe	
Hall coefficient R [a]	-1.2×10^{-10}		0.34×10^{-10}		0.69×10^{-6}			$m^3/(A\,s)$	
Thermoelectronic coefficient			5.0					μV/K	
Electronic work function	3.67		3.47					V	
Thermal work function	3.42	4.8	3.47					V	
Molar magnetic susceptibility χ_{mol} (SI)	1.22×10^{-3}	3.48×10^{-3}	5.14×10^{-3}	7.23×10^{-3}	6.60×10^{-3}			cm^3/mol	
Molar magnetic susceptibility χ_{mol} (cgs)	97×10^{-6}	277×10^{-6}	409×10^{-6}	575×10^{-6}	525×10^{-6}			cm^3/mol	
Mass magnetic susceptibility χ_{mol} (SI)	7.2×10^{-6}		21.6×10^{-6}		31.7×10^{-6}	50×10^{-6}		cm^3/g	

[a] At 298 K, $B = 0.3–0.7$ T.

Element name	Berkelium	Californium	Einsteinium	Fermium	Mendelevium	Nobelium	Lawrencium	
Chemical symbol	**Bk**	**Cf**	**Es**	**Fm**	**Md**	**No**	**Lr**	
Atomic number Z	**97**	**98**	**99**	**100**	**101**	**102**	**103**	
Characteristics	Metal	Metal	Metal		Metal			

Table 2.1–26C Actinides. Allotropic and high-pressure modifications

Element	Thorium		Protactinium		Uranium			
Modification	α-Th	β-Th	α-Pa	β-Pa	α-U	β-U	γ-U	Units
Crystal system, Bravais lattice	cub, fc	cub, bc	tetr, I	cub, bc	orth, C	tetr	cub, bc	
Structure type	Cu	W	In	W	α-U		W	
Lattice constant *a*	508.51	411	394.5		285.38	1075.9	352.4	pm
Lattice constant *b*					586.80			pm
Lattice constant *c*			324.2		495.57	565.4		pm
Space group	$Fm\bar{3}m$	$Im\bar{3}m$	$I4/mmm$	$Im\bar{3}m$	$Cmcm$	$P4_2/mmm$	$Im\bar{3}m$	
Schoenflies symbol	O_h^5	O_h^9	D_{4h}^{17}	O_h^9	D_{2h}^{17}	C_{2h}^{14}	O_h^9	
Strukturbericht type	A1	A2	A6	A2	A20		A2	
Pearson symbol	cF4	cI2	tI2	cI2	oC4	tP30	cI2	
Number *A* of atoms per cell	4	2	2	2	4	30	2	
Coordination number	12	8	8+2	2+2+4+4	2+2+4+4	12	8	
Shortest interatomic distance, solid	360	356	321		276	287–353	347	pm
Range of stability	RTP	> 1673 K	RTP	> 1443 K	RTP	> 935 K	> 1045 K	

Element	Neptunium			Plutonium						
Modification	α-Np	β-Np	γ-Np	α-Pu	β-Pu	γ-Pu	δ-Pu	δ′-Pu	ε-Pu	Units
Crystal system, Bravais lattice	orth, P	tetr	cub, bc	mon, P	mon	orth	cub, fc	tetr	cub, bc	
Structure type	α-Np		W				Cu	In	W	
Lattice constant *a*	666.3	489.6	352	618.3	928.4	315.87	463.71	332.61	570.3	pm
Lattice constant *b*	472.3			482.2	1046.3	576.82				pm
Lattice constant *c*	488.7	338.7		1096.8	785.9	1016.2		446.30		pm
Lattice angle *γ*				101.78	92.13					deg
Space group	$Pnma$	$P42_12$	$Im\bar{3}m$	$P2_1/m$	$I2/m$	$Fddd$	$Fm\bar{3}m$	$I4/mmm$	$Im\bar{3}m$	
Schoenflies symbol	D_{2h}^{16}	D_{4h}^{7}	O_h^9	C_{2h}^{2}	C_{2h}^{3}	D_{2h}^{24}	O_h^5	D_{4h}^{17}	O_h^9	
Strukturbericht type	A_c	A_d	A2				A1	A6	A2	
Pearson symbol	oP8	tP4	cI2	mP16	mI34	oF8	cF4	tI2	cI2	
Number *A* of atoms per cell	8	4	2	16	34	8	4	2	2	
Coordination number	8+6	8+6	8+6			4+2+4	12	4+8	8+6	
Shortest interatomic distance, solid	260–336	275–356	305	257–278	259–310	303	328	333	315	pm
Range of stability	RTP	>553 K	>850 K	RTP	>395 K	>508 K	>592 K	>726 K	>744 K	

Table 2.1-26C Actinides. Allotropic and high-pressure modifications, cont.

Element	Americium			Curium		
Modification	α-Am	β-Am	γ-Am	α-Cm	β-Cm	
						Units
Crystal system, Bravais lattice	hex	cub, fc	orth	hex	cub, fc	
Structure type	α-La	Cu	α-U	α-La	Cu	
Lattice constant *a*	346.8	489.4	306.3	349.6	438.1	pm
Lattice constant *b*			596.8			pm
Lattice constant *c*	1124.1		516.9	1133.1		pm
Space group	$P6_3/mmc$	$Fm\bar{3}m$	$Cmcm$	$P6_3/mmc$	$Fm\bar{3}m$	
Schoenflies symbol	D^4_{6h}	O^5_h	D^{17}_{2h}	D^4_{6h}	O^5_h	
Strukturbericht type	A3′	A1	A20	A3′	A1	
Pearson symbol	hP4	cF4	oC4	hP4	cF4	
Number *A* of atoms per cell	4	4	4	4	4	
Coordination number	6+6	12		6+6	12	
Shortest interatomic distance, solid	347	346		350	310	pm
Range of stability	RTP	> 878 K	> 15.0 GPa	RTP	> 1449 K	

Element	Berkelium		Californium		Einsteinium		
Modification	α-Bk	β-Bk	α-Cf	β-Cf	α-Es	β-Es	
							Units
Crystal system, Bravais lattice	hex	cub, fc	hex	cub, fc	hex	cub, fc	
Structure type	α-La	Cu	α-La	Cu	α-La	Cu	
Lattice constant *a*	341.6	499.7					pm
Lattice constant *c*	1106.9						pm
Space group	$P6_3/mmc$	$Fm\bar{3}m$	$P6_3/mmc$	$Fm\bar{3}m$	$P6_3/mmc$	$Fm\bar{3}m$	
Schoenflies symbol	D^4_{6h}	O^5_h	D^4_{6h}	O^5_h	D^4_{6h}	O^5_h	
Strukturbericht type	A3′	A1	A3′	A1	A3′	A1	
Pearson symbol	hP4	cF4	hP4	cF4	hP4	cF4	
Number *A* of atoms per cell	4	4	4	4	4	4	
Coordination number		12		12			
Shortest interatomic distance, solid		353		406			pm
Range of stability	RTP	> 1183 K		> 1213 K	RTP	> 1093 K	

Table 2.1-26D Actinides. Ionic radii (determined from crystal structures)

Element	Thorium	Protactinium			Uranium				Neptunium				
Ion	Th^{4+}	Pa^{3+}	Pa^{4+}	Pa^{5+}	U^{3+}	U^{4+}	U^{5+}	U^{6+}	Np^{3+}	Np^{4+}	Np^{5+}	Np^{6+}	
Coordination number													Units
2								45					pm
4								52					pm
6	94	104	90	78	103	89	76	73	101	87	75	72	pm
8	105					100		86					pm
10	113												pm
12	121					117							pm

Table 2.1-26D Actinides. Ionic radii (determined from crystal structures), cont.

Element	Plutonium				Americium		Curium		
Ion	Pu^{3+}	Pu^{4+}	Pu^{5+}	Pu^{6+}	Am^{3+}	Am^{4+}	Cm^{3+}	Cm^{4+}	
Coordination number									Units
6	100	86	74	71	98	85	97	85	pm
8					109	95		95	pm

Element	Berkelium		Californium		
Ion	Bk^{3+}	Bk^{4+}	Cf^{3+}	Cf^{4+}	
Coordination number					Units
6	96	83	95	82	pm
8		93		92	pm

References

1.1 W. Martienssen (Ed.): *Numerical Data and Functional Relationships in Science and Technology*, Landolt-Börnstein, New Series III and IV (Springer, Berlin, Heidelberg 1970–2003)

1.2 R. Blachnik (Ed.): *Elemente, anorganische Verbindungen und Materialien, Minerale*, D'Ans-Lax, Taschenbuch für Chemiker und Physiker, Vol. 3, 4th edn. (Springer, Berlin, Heidelberg 1998)

1.3 D. R. Lide (Ed.): *CRC Handbook of Chemistry and Physics*, 80th edn. (CRC Press, Boca Raton 1999)

1.4 Lehrstuhl für Werkstoffchemie, T.H. Aachen: *Thermodynamic Properties of Inorganic Materials*, Landolt-Börnstein, New Series IV/19 (Springer, Berlin, Heidelberg 1999)

Subject Index

π-bonded chain geometry 993
π-bonded chain model
– diamond(111)2×1 1005
$\alpha = 0$ glass-ceramics 558
$(CH_3NHCH_2COOH)_3 \cdot CaCl_2$ family 932
5CB
– liquid crystals 948
8CB
– liquid crystals 948
8OCB
– liquid crystals 949

A

Abbe value
– glasses 543, 548
Abrikosov vortices 717
absorption and fluorescence spectra of CdSe 1039
absorption coefficient
– two-photon 826
absorption spectra of spherical particles 1046
Ac actinium 84
acceptor surface level 1023
acceptor surface state 1022
accumulation layer 1020
accuracy 4
acids
– liquid crystals 946
acoustic band 1012
acoustic surface wave 906
acronyms
– solid surface 1026
actinium Ac
– elements 84
adatom 995
adopted numerical values for selected quantities 23
Ag silver 65
Ag-based materials 344
age hardening 198
AISI (American Iron and Steel Institute) 221
Al aluminium 78
Al bronzes 298
alkali aluminium silicates
– electrical properties 434
– mechanical properties 434
– thermal properties 434
alkali halides
– surface phonon energy 1017
alkali–alkaline-earth silicate glasses 530
alkali–lead silicate glasses 530
alkaline-earth aluminium silicates
– electrical properties 436
– glasses 530
– mechanical properties 435
– thermal properties 436
allotropic and high-pressure modifications
– elements 46
alloy
– cast irons 270
– cobalt 272
– elinvar 780
– invar 780
– lead, battery grid 413
– lead–antimony 412
– lead–tin 415
– magnesium 163
– Ti_3Al-based 210
– TiAl-based 213
– titanium 206
– wear resistant 274
alloy systems 296
Allred 46
Alnico 798
Al–O–N ceramics
– dielectric properties 447
– optical properties 447
alum family 911
– ferroelectrics 911
alumina
– electrical properties 446
– mechanical properties 445
– properties 445
– thermal properties 446
aluminium Al
– aluminium alloys 171
– aluminium production 171
– chemical properties 172
– cold working 195
– corrosion behavior 204
– elements 78
– hot working 195
– mechanical properties 172
– mechanical treatment 195
– surface layers 204
– work hardening 195
aluminium alloy
– abrasion resistance 190
– aging 198
– behavior in magnetic fields 194
– binary Al-based systems 174
– classification of aluminium alloys 179
– coefficient of thermal expansion 192
– creep behavior 187
– elastic properties 194
– electrical conductivity 194
– hardness 186
– homogenization 198
– machinability 192
– mechanical properties 180, 182
– nuclear properties 194
– optical properties 194
– physical properties 192, 193
– sheet formability 190
– soft annealing 197
– specific heat 194
– stabilization 197
– stress-relieving 198
– structure 182
– technical property 186
– technological properties 190
– tensile strength 186
– thermal softening 195
– work-hardenable 180
aluminium antimonide
– crystal structure, mechanical and thermal properties 610
– electromagnetic and optical properties 619
– electronic properties 616
– transport properties 618
aluminium arsenide
– crystal structure, mechanical and thermal properties 610
– electromagnetic and optical properties 619
– electronic properties 616
– transport properties 618
aluminium casting alloys
– mechanical properties 184
– structure 184
aluminium compounds
– crystal structure, mechanical and thermal properties 610

– electromagnetic and optical properties 619
– electronic properties 614
– mechanical properties 610
– phonon dispersion curves 612
– thermal conductivity 618
– thermal properties 610
– transport properties 617
aluminium nitride
– crystal structure, mechanical and thermal properties 610
– electromagnetic and optical properties 619
– electronic properties 616
– transport properties 618
aluminium phase diagram
– aluminium alloy phase diagram 174
aluminium phosphide
– crystal structure, mechanical and thermal properties 610
– electromagnetic and optical properties 619
– electronic properties 616
– transport properties 618
aluminothermy 174
Al–Ni phase diagram 285
Am americium 151
americium Am
– elements 151
amorphous alloys
– cobalt–based 774
– iron–based 773
– nickel–based 774
amorphous materials 27, 39
amorphous metallic alloys 772
amount of substance
– definition 14
ampere
– SI base unit 14
amphiphilic compound 941
amphiphilic liquid crystal 942
anisotropic magnetoresistance 1058
annealing coefficient
– glasses 548
annealing of steel 223
antiferroelectric crystal 903
antiferroelectric hysteresis loop 904
antiferroelectric liquid crystal 911
antiferroelectrics
– definition 903
– dielectric properties 903
– elastic properties 903
– pyroelectric properties 903
antimony Sb
– elements 98
aperiodic crystals 27
aperiodic materials 33
apparent tilt angle 935
Ar argon 128
area of surface primitive cell
– crystallographic formulas 986
argon Ar
– elements 128
arsenic As
– elements 98
ARUPS 997
As arsenic 98
astatine At
– elements 118
ASTM 241, 242
ASTM (American Society for Testing and Materials) 330
ASW 906
At astatine 118
atomic moment 755
atomic number *Z*
– elements 45
atomic radius
– elements 46
atomic scattering 1019
atomic, ionic, and molecular properties
– elements 46
atomically clean crystalline surface 979
atom–surface potential
– surface phonons 1019–1020
Au gold 65
austenitizing 224

B

B boron 78
Ba barium 68
back-bond state 1006
bainite 223
$BaMnF_4$ family 922
band bending
– solid surfaces 1023
band gap *see* energy gap 592
band pass filters
– glasses 566
band structure
– aluminium compounds 614
– beryllium compounds 653
– boron compounds 606
– cadmium compounds 679
– group IV semiconductors and IV–IV compounds 589–592
– indium compounds 643
– magnesium compounds 657
– mercury compounds 688
– oxides of Ca, Sr, and Ba 662
– zinc compounds 668
barium Ba
– elements 68
barium oxide
– crystal structure, mechanical and thermal properties 660
– electromagnetic and optical properties 664
– electronic properties 661
– transport properties 663
barium titanate 915
base quantities 12
– ISO 13
base unit
– SI 13
basis
– crystal structure 28
$BaTiO_3$ 915
bcc positions
– surface diagrams 982
Be beryllium 68
becquerel
– SI unit of activity 19
benzene 946
berkelium Bk
– elements 151
beryllium Be
– elements 68
beryllium compounds 652
– crystal structure, mechanical and thermal properties 652
– electromagnetic and optical properties 655
– electronic properties 653
– mechanical and thermal properties 652
– optical properties 655
– thermal properties 652
– transport properties 655
beryllium oxide 447
– crystal structure, mechanical and thermal properties 652
– electrical properties 448
– electronic properties 653
– mechanical properties 447
– thermal properties 448
beryllium selenide
– crystal structure, mechanical and thermal properties 652
– electronic properties 653
beryllium sulfide
– crystal structure, mechanical and thermal properties 652

Subject Index

– electronic properties 653
beryllium telluride
– crystal structure, mechanical and thermal properties 652
– electronic properties 653
Bethe–Slater–Pauling relation 755, 756
Bh Bohrium 124
Bi bismuth 98
biaxial crystals 826
binding energy 998
– metal 999
Bioverit
– glasses 559
BIPM (Bureau International des Poids et Mesures) 3, 11, 12
bismuth Bi
– elements 98
Bi – Sr – Ca – Cu – O (BSCCO) 736
Bi–Sr–Ca–Cu–O
– coherence lengths 736
– London penetration depths 736
– maximum T_c 736
– structural data 736
– superconducting properties 741
– upper critical fields 736
BK 7
– glasses 537
Bk berkelium 151
blue phase 942
BNN
– ferroelectric material 920
Bohrium Bh
– elements 124
boiling temperature
– elements 47
Bondi 46
boracite-type family 911, 921
borides
– physical properties 452
Borofloat
– glasses 528, 529
boron antimonide
– crystal structure, mechanical, and thermal properties 604
– electromagnetic and optical properties 610
– electronic properties 607
– transport properties 608
boron arsenide
– crystal structure, mechanical, and thermal properties 604
– electromagnetic and optical properties 610
– electronic properties 606
– transport properties 608
boron B
– elements 78
boron compounds
– crystal structure, mechanical and thermal properties 604
– electromagnetic and properties 610
– electronic properties 606
– mechanical properties 604
– thermal properties 604
– transport properties 608
boron nitride
– crystal structure, mechanical, and thermal properties 604
– electromagnetic and optical properties 610
– electronic properties 606
– transport properties 608
boron phosphide
– crystal structure, mechanical, and thermal properties 604
– electromagnetic and optical properties 610
– electronic properties 606
– transport properties 608
borosilicate glasses 529, 530
Br bromine 118
Bragg equation 40
brasses 298
Bravais cell 979
– 2D lattice 980
Bravais lattice 32
– elements 47
breathing-mode acoustic oscillations 1040
Brillouin scattering 906
Brillouin zone 913
– aluminium compounds 614
– beryllium compounds 653
– boron compounds 606
– cadmium compounds 678
– gallium compounds 626
– group IV semiconductors and IV–IV compounds 589
– indium compounds 643
– magnesium compounds 657
– mercury compounds 688
– oxides of Ca, Sr, and Ba 661
– zinc compounds 668
Brillouin zone corner 921
bromine Br
– elements 118
bronzes 298
BSCCO
– films 738
– single crystal 739
– tapes 739
– wires 739
buckled dimer 991
bulk electron density 998
bulk glassy alloys 217, 218
bulk mobility 1026
bulk modulus
– elements 47

C

C carbon 88
Ca calcium 68
cadmium Cd
– elements 73
cadmium compounds
– crystal structure, mechanical and thermal properties 676
– electromagnetic and optical properties 683
– electronic properties 678
– mechanical and thermal properties 676
– optical properties 683
– thermal properties 676
– transport properties 682
cadmium oxide
– crystal structure, mechanical and thermal properties 676
– electromagnetic and optical properties 683
– electronic properties 678
– transport properties 682
cadmium selenide
– crystal structure, mechanical and thermal properties 676
– electromagnetic and optical properties 683
– electronic properties 678
– transport properties 682
cadmium sulfide
– crystal structure, mechanical and thermal properties 676
– electromagnetic and optical properties 683
– electronic properties 678
– transport properties 682
cadmium telluride
– crystal structure, mechanical and thermal properties 676
– electromagnetic and optical properties 683
– electronic properties 678
– transport properties 682
calamitic liquid crystal 941, 942

calcium Ca
– elements 68
calcium oxide
– crystal structure, mechanical and thermal properties 660
– electromagnetic and optical properties 664
– electronic properties 661
– transport properties 663
californium Cf
– elements 151
candela
– SI base unit 15
capacitor 903
capillary viscometer 943
carat 23
carbide
– cemented 277
– electrical properties 466
– mechanical properties 466
– physical properties 458
– thermal properties 466
carbon C
– elements 88
carbon equivalent (CE) 268
carbon fibers
– physical properties 477
carbon steels 230
carrier concentration n_i
– gallium compounds 631
– indium compounds 647
cast
– classification 268
cast iron 268
– grades 268
– mechanical properties 270
casting technology 170
catalysis 1020
Cd cadmium 73
Ce cerium 142
cell surface 943
cellulose 481, 509
– cellulose acetate (CA) 509, 510
– cellulose acetobutyrate (CAB) 509, 510
– cellulose propionate (CP) 509, 510
– ethylcellulose (EC) 509, 510
– polymers 509
– vulcanized fiber (VF) 509, 510
cement 432
cemented carbides 277
centering types
– crystal structure 28
centrosymmetric media 1007
Cerabone
– glasses 559
ceramic capacitor 906
ceramic thin film 914
– ferroelectrics 903
ceramics 345, 431
– Al–O–N 447
– applications 432
– non-oxide 451
– oxide 437
– properties 432
– refractory 437
– silicon 433
– technical 437
– traditional 432
Ceran®
– glasses 559
– linear thermal expansion 558
Ceravital
– glasses 559
cerium Ce
– elements 142
cesium Cs
– elements 59
Cf californium 151
CGPM (Conférence Générale des Poids et Mesures) 11, 12
CGS
– electromagnetic system 21
– electrostatic system 21
– Gaussian system 21
cgs definitions of magnetic susceptibility 48
chalcogenide glasses 568, 571
channel conductivity 1020
characterization of optical glasses 547
Charpy impact strength 478
– polymers 478
chemical disorder 38
chemical stability
– glasses 530, 531
– optical glasses 550
chemical symbols
– element 45
chemisorption 1020
chiral smectic C
– liquid crystals 944
chlorine Cl
– elements 118
cholesteric phase 942
cholesteryl (cholest-5-ene) substituted mesogens
– liquid crystals 968
cholesteryl compound 944
chromium Cr
– elements 114
CIP (current in the plane of layer) 1051
CIPM (Comité International des Poids et Mesures) 3, 11
Cl chlorine 118
clamped crystal 907, 921, 934
clamped dielectric constant 907
classifications of liquid crystals 942
climatic influences
– glasses 550
cluster boundaries 907
cluster formation 907
CM
– liquid crystals 970
Cm curium 151
CMOS (complementary metal-oxide-semiconductor) 1059
Co cobalt 135
$Co_{17}RE_2$ 805
Co_5RE 805
coating 943
cobalt
– alloys 272
– applications 273
– hard-facing alloy 274
– mechanical properties 277
– superalloys 274
– surgical implant alloys 277
cobalt Co
– elements 135
cobalt corrosion-resistant alloys 276
cobalt-based corrosion-resistant alloys 276
CODATA (Committee on Data for Science and Technology) 4
coefficient of expansion 478
– polymers 478
coefficient of thermal expansion
– glasses 526
coercive field 904
coherence length 920
coherent phonon and Raman spectra 1041
colloidal synthesis
– nanostructured materials 1065
colored glasses
– colorants 567
– glasses 565
columnar phase 941
commensurate reconstruction 986
commercially pure titanium (cp-Ti) 206
communication technology 912

compacted (vermicular) graphite (CG) 268
complex perovskite-type oxide 909
complex refractive index
– zinc compounds 674
composite medium 1045
– dielectric constant 1045
composite solder glasses
– glasses 563
composite structures
– crystallography 34
compound semiconductor 1003
compressibility 478
– polymers 478
compression modulus
– elements 47
condensed matter 27
– classification 28
conductivity
– frequency-dependent 823
conductivity tensor
– elements 47
conductor
– nanoparticle-based 1043
confined electronic systems
– nanostructured materials 1035
confinement effect
– nanostructured materials 1031
constants
– fundamental 3
container glasses 529
continuous distribution of states 1022
continuous-cooling-transformation (CCT) diagram 238
controlled rolling 240
Convention du Mètre 12
conventional system
– ISO 13
conversion factor 945
– density 945
– diamagnetic anisotropy 945
– dipole moment 945
– dynamic viscosity 945
– kinematic viscosity 945
– molar mass 945
– temperatures of phase transitions 945
– thermal conductivity 945
cooperative interactions 903
coordination number
– elements 46, 47
copolymers
– physical properties 477, 483
copper
– unalloyed 297
copper alloys 296, 297
copper Cu
– elements 65
copper–nickel 300
copper–nickel–zinc 300
Co–RE
– phase equilibria 805
corrosion 1020
– resistance 218
Coulomb blockade
– nanostructured materials 1031, 1044
coupled plasmon modes 1048
Co–Sm 803
CPP (current flows perpendicular to the plane of the layer) 1051, 1054
Cr chromium 114
creep modulus 478
– polymers 478
critical field 904
critical slowing-down 907, 908
critical temperature
– elements 47, 48
– Pb alloys 699
CrNi steels 252
Cronstedt, swedish mineralogist 1032
crystal axes 980
crystal morphology 30
crystal optics 824
crystal structure
– beryllium compounds 652
– cadmium compounds 676
– group IV semiconductors 578
– III–V compounds 610
– indium compounds 638
– IV–IV compound semiconductors 578
– magnesium compounds 655
– mercury compounds 686
– oxides of Ca, Sr, and Ba 660
– zinc compounds 665
crystal structure, mechanical and thermal properties
– III–V compounds 621, 638
– III–V semiconductors 604
– II–VI compounds 652, 655, 660, 665, 676, 686
crystal symmetry
– elements 47
– ferroelectrics 915
crystalline ferroelectric 911
crystalline materials
– definition 27
crystalline surface
– atomically clean 979
crystallization
– glasses 524
crystallographic formulas 986
crystallographic properties
– elements 46, 47
crystallographic space group 946–952, 955–961, 964, 966, 968–971
crystallographic structure
– methods to investigate 39
crystallography
– concepts and terms 27
– rudiments of 27
crystals
– biaxial 826
– cubic 826
– isotropic 826
– uniaxial 826
Cryston
– glasses 559
Cs cesium 59
CS2004
– liquid crystals 975
Cu copper 65
cubic
– dielectrics 828, 838
cubic $BaTiO_3$ 916
cubic boron nitride 451
cubic crystal 47
cubic system 986
Curie point 906
Curie temperature 906
– surface 1009
Curie–Weiss constant 931, 932
curium Cm
– elements 151
current/voltage characteristics of films 1044
Cu–Ni
– electrical conductivity 302
– thermal conductivity 303
Cu–Zn
– electrical conductivity 302
– thermal conductivity 303
Cu–Zn phase diagram 299
cyclohexane 946
cycloolefine copolymer (COC) 486, 487
cyclosilicate 433

D

D deuterium 54
damping constant
– optical mode frequency 916
dangling bonds (DBs) 991

DAS 991
data storage media
– nanostructured materials 1031
Db dubnium 105
de Broglie wavelength 1035
Debye length
– solid surfaces 1020
Debye temperature Θ_D
– boron compounds 605
– cadmium compounds 678
– gallium compounds 623
– group IV semiconductors 584
– indium compounds 640
– IV–IV compound semiconductors 584
– mercury compounds 687
– metal surfaces 1013
– oxides of Ca, Sr, and Ba 661
– solid surfaces 1014
– surface phonons 1012
– zinc compounds 667
decimal multiples of SI units 19
degree Celsius
– unit of temperature 14
density ϱ 478, 945–954, 956–971
– aluminium compounds 611
– beryllium compounds 653
– boron compounds 604
– cadmium compounds 676
– elements 47
– gallium compounds 621
– group IV semiconductors and IV–IV compounds 579
– indium compounds 638
– magnesium compounds 656
– mercury compounds 686
– oxides of Ca, Sr, and Ba 660
– polymers 478
– temperature dependence 943
– zinc compounds 665
density of electronic states
– magnesium compounds 658
density of electronic states *see also* density of states 658
density of phonon states *see also* density of states 658
density of states (DOS)
– nanostructure 1034
dentistry 330
depletion layer 1020
– solid surfaces 1024
derived quantities 12
– ISO 13
derived units
– SI 16
– special names and symbols 16
deuteration 925
deuterium D
– elements 54
devices
– optical 903
– piezoelectric 903
– pyroelectric 903
devil's staircase 932
devitrifying solder glasses 562
DFB (distributed feedback) 1038
DFG (difference frequency generation) 825
diamagnetic anisotropy 943, 945
diamond
– crystal structure, mechanical and thermal properties 578–588
– electromagnetic and optical properties 601
– electronic properties 589–594
– transport properties 595
diamond positions
– surface diagrams 983
diamond-like structure 979
Dicor
– glasses 559
dielectric
– constant (low-frequency) 823
– dissipation factor 823
– elasticity 823
– general properties 822
– lossy 823
– low-frequency materials 822
– stiffness constant 823
dielectric anisotropy 943, 946, 947
dielectric anomaly 922, 934
dielectric constant ε 907, 947–958, 960–966, 968–971, 973, 1045
– aluminium compounds 619
– beryllium compounds 655
– boron compounds 610
– cadmium compounds 683
– elements 46
– gallium compounds 635, 637
– group IV semiconductors and IV–IV compounds 601–603
– indium compounds 650
– magnesium compounds 660
– mercury compounds 691
– oxides of Ca, Sr, and Ba 664
– zinc compounds 672
dielectric dispersion 907
dielectric dissipation factor
– glasses 538
dielectric function 1001
– surface layer 1007
dielectric loss tan δ
– gallium compounds 635
dielectric material
– properties 826
dielectric polarization 825
dielectric properties
– glasses 538
dielectric strength
– glasses 539
dielectric tensor 827
dielectrics
– α-iodic acid, α-HIO_3 868
– α-mercuric sulfide, α-HgS 846
– α-silicon carbide, SiC 840
– α-silicon dioxide, α-SiO_2 846
– α-zinc sulfide, α-ZnS 842
– β-barium borate, β-BaB_2O_4 848
– 2-cyclooctylamino-5-nitropyridine, $C_{13}H_{19}N_3O_2$ 874
– 3-methyl 4-nitropyridine 1-oxide, $C_6N_2O_3H_6$ 868
– 3-nitrobenzenamine, $C_6H_4(NO_2)NH_2$ 878
– 4-($N1N$-dimethylamino)-3-acetamidonitrobenzene 884
– ADA 856
– ADP 856
– aluminium oxide, α-Al_2O_3 844
– aluminium phosphate $AlPO_4$ 844
– ammonium dideuterium phosphate, $ND_4D_2PO_4$ 856
– ammonium dihydrogen arsenate, $NH_4H_2AsO_4$ 856
– ammonium dihydrogen phosphate, $NH_4H_2PO_4$ 856
– ammonium Rochelle salt 870
– ammonium sulfate, $(NH_4)_2SO_4$ 872
– "banana" 872
– barium fluoride, BaF_2 828
– barium formate, $Ba(COOH)_2$ 866
– barium magnesium fluoride, $BaMgF_4$ 872
– barium nitrite monohydrate, $Ba(NO_2)_2 \cdot H_2O$ 842
– barium sodium niobate, $Ba_2NaNb_5O_{15}$ 872
– Barium titanate, $BaTiO_3$ 864
– BBO 848
– berlinite 844
– beryllium oxide, BeO 840
– BGO 838
– BIBO 884
– bismuth germanium oxide, $Bi_{12}GeO_{20}$ 838

– bismuth silicon oxide, $Bi_{12}SiO_{20}$ 838
– bismuth triborate, BiB_3O_6 884
– BK7 Schott glass 828
– BMF 872
– BSO 838
– cadmium germanium arsenide, $CdGeAs_2$ 856
– cadmium germanium phosphide, $CdGeP_2$ 856
– cadmium selenide, CdSe 840
– cadmium sulfide, CdS 840
– cadmium telluride, CdTe 834
– calcite, $CaCO_3$ 844
– calcium fluoride, CaF_2 828
– calcium tartrate tetrahydrate, $Ca(C_4H_4O_6)\cdot 4H_2O$ 874
– CBO 868
– CDA 858
– cesium dideuterium arsenate, CsD_2AsO_4 856
– cesium dihydrogen arsenate, CsH_2AsO_4 858
– cesium lithium borate, $CsLiB_6O_{10}$ 858
– cesium triborate, CsB_3O_5 868
– cinnabar 846
– CLBO 858
– CNB 874, 875
– COANP 874
– copper bromide, CuBr 834
– copper chloride, CuCl 834
– copper gallium selenide, $CuGaSe_2$ 858
– copper gallium sulfide, $CuGaS_2$ 858
– copper iodide, CuI 834
– cubic $m3m$ (O_h) 828
– cubic $\bar{4}3m$ (T_d) 834
– cubic, 23 (T) 838
– D(+)-saccharose, $C_{12}H_{22}O_{11}$ 888
– DADP 856
– DAN 884
– DCDA 856
– deuterated L-arginine phosphate, $(ND_xH_{2-x})_2^+(CND)(CH_2)_3CH$ $(ND_yH_{3-y})^+COO^-\cdot D_2PO_4^- \cdot D_2O$ 884
– diamond, C 830
– dipotassium tartrate hemihydrate, $K_2C_4H_4O_6\cdot 0.5H_2O$ 886
– DKDA 858
– DKDP 858
– DKT 886
– DLAP 884, 886
– DRDA 860
– DRDP 860
– fluorite 828
– fluorspar 828
– forsterite 866
– gadolinium molybdate, $Gd_2(MoO_4)_3$ 876
– gallium antimonide, GaSb 834
– gallium arsenide, GaAs 834
– gallium nitride, GaN 840
– gallium phosphide, GaP 834
– gallium selenide, GaSe 838
– gallium sulfide, GaS 838
– germanium, Ge 830
– GMO 876
– greenockite 840
– halite 832
– hexagonal, 6 (C_6) 842
– hexagonal, $6mm$ (C_{6v}) 840
– hexagonal, $\bar{6}m2$ (D_{3h}) 838
– high-frequency (optical) properties 817
– Iceland spar 844
– indium antimonide, InSb 836
– indium arsenide, InAs 836
– indium phosphide, InP 836
– Irtran-3 828
– Irtran-6 834
– isotropic 828
– KB5 880
– KBBF 846
– KDA 860
– KDP 860
– KLINBO 864
– KTA 880
– KTP 880
– LBO 878
– L-CTT 874
– lead molybdate, $PbMoO_4$ 854
– lead titanate, $PbTiO_3$ 864
– LFM 876
– list of described substances 818
– lithium fluoride, LiF 830
– lithium formate monohydrate, $LiCOOH\cdot H_2O$ 876
– lithium gallium oxide, $LiGaO_2$ 876
– lithium iodate, α-$LiIO_3$ 842
– lithium metagallate 876
– lithium niobate (5% MgO-doped), MgO:$LiNbO_3$ 848
– lithium niobate, $LiNbO_3$ 848
– lithium sulfate monohydrate, $Li_2SO_4\cdot H_2O$ 886
– lithium tantalate, $LiTaO_3$ 848
– lithium tetraborate, $Li_2B_4O_7$ 864
– lithium triborate, LiB_3O_5 878
– low-frequency properties 817
– magnesium fluoride, MgF_2 852
– magnesium oxide, MgO 830
– magnesium silicate, Mg_2SiO_4 866
– Maxwell's equations 824
– m-chloronitrobenzene, $ClC_6H_4NO_2$ 874, 875
– mNA 878
– m-nitroaniline 878
– MNMA 874
– monoclinic, 2 (C_2) 884
– N,2-dimethyl-4-nitrobenzenamine, $C_8H_{10}N_2O_2$ 874
– N-[2-(dimethylamino)-5-nitrophenyl]-acetamide 884
– nantokite 834
– numerical data 818, 828
– orthorhombic, 222 (D_2) 866
– orthorhombic, $mm2$ (C_{2v}) 872
– orthorhombic, mmm (D_{2h}) 866
– paratellurite 854
– physical properties 817
– PMMA (Plexiglas) 828
– POM 868
– potassium acid phthalate, $KH(C_8H_4O_4)$ 878
– potassium bromide, KBr 830
– potassium chloride, KCl 830
– potassium dideuterium arsenate, KD_2AsO_4 858
– potassium dideuterium phosphate, KD_2PO_4 858
– potassium dihydrogen arsenate, KH_2AsO_4 860
– potassium dihydrogen phosphate, KH_2PO_4 860
– potassium fluoroboratoberyllate, $KBe_2BO_3F_2$ 846
– potassium iodide, KI 830
– potassium lithium niobate, $K_3Li_2Nb_5O_{15}$ 864
– potassium niobate, $KNbO_3$ 880
– potassium pentaborate tetrahydrate, $KB_5O_8\cdot 4H_2O$ 880
– potassium sodium tartrate tetrahydrate, $KNa(C_4H_4O_6)\cdot 4H_2O$ 870
– potassium titanate (titanyl) phosphate, $KTiOPO_4$ 880
– potassium titanyl arsenate, $KTiOAsO_4$ 880
– proustite 850
– pyragyrite 850
– quartz 846

– RDA 860
– RDP 862
– Rochelle salt 870
– rock salt 832
– RTP 882
– rubidium dideuterium arsenate, RbD_2AsO_4 860
– rubidium dideuterium phosphate, RbD_2PO_4 860
– rubidium dihydrogen arsenate, RbH_2AsO_4 860
– rubidium dihydrogen phosphate, RbH_2PO_4 862
– rubidium titanate (titanyl) phosphate, $RbTiOPO_4$ 882
– rutile 852
– sapphire 844
– Silicon dioxide, SiO_2 828
– silicon, Si 830
– silver antimony sulfide, Ag_3SbS_3 850
– silver arsenic sulfide, Ag_3AsS_3 850
– silver gallium selenide, $AgGaSe_2$ 862
– silver thiogallate, $AgGaS_2$ 862
– sodium ammonium tartrate tetrahydrate, $Na(NH_4)C_4H_4O_6 \cdot 4H_2O$ 870
– sodium chlorate, $NaClO_3$ 838
– sodium chloride, NaCl 832
– sodium fluoride, NaF 832
– sodium nitrite, $NaNO_2$ 882
– strontium fluoride, SrF_2 832
– strontium titanate, $SrTiO_3$ 832
– sucrose 888
– sylvine 830
– sylvite 830
– TAS 850
– tellurium dioxide, TeO_2 854
– tellurium, Te 846
– tetragonal, $4/m$ (C_{4h}) 854
– tetragonal, $4/mmm$ (D_{4h}) 852
– tetragonal, 422 (D_4) 854
– tetragonal, $4mm$ (C_{4v}) 864
– tetragonal, $\bar{4}2m$ (D_{2d}) 856
– TGS 888
– thallium arsenic selenide, Tl_3AsSe_3 850
– titanium dioxide, TiO_2 852
– tourmaline, $(Na,Ca)(Mg,Fe)_3B_3Al_6Si_6(O,OH,F)_{31}$ 850
– triglycine sulfate, $(CH_2NH_2COOH)_3 \cdot H_2SO_4$ 888
– trigonal, 32 (D_3) 844
– trigonal, $3m$ (C_{3v}) 848
– trigonal, $\bar{3}m$ (D_{3d}) 844
– urea, $(NH_2)_2CO$ 862
– wurtzite 842
– YAG 832
– YAP 866
– YLF 854
– yttrium aluminate, $YAlO_3$ 866
– yttrium aluminium garnet, $Y_3Al_5O_{12}$ 832
– yttrium lithium fluoride, $YLiF_4$ 854
– yttrium vanadate, YVO_4 852
– YVO 852
– zinc blende 842
– zinc germanium diphosphide, $ZnGeP_2$ 862
– zinc oxide, ZnO 840
– zinc selenide, ZnSe 836
– zinc telluride, ZnTe 836
– zincite 840
difference frequency generation (DFG) 825
differential conductance as function of voltage 1043
diffraction method 39
diffusion coefficient 946–949, 956–960, 970
diffusion-controlled growth
– nanostructured materials 1064
dimensionality
– nanostructured materials 1033
dimensions
– physical quantities 4, 13
dimer bond length 995
dioxide
– zirconium 448
dipole glass 906
dipole moment 945, 947–953, 955–963, 968–971
direct gap
– group IV semiconductors and IV–IV compounds 592
direct piezoelectric effect 824
director 943
discotic liquid crystal 941, 942, 972
– physical properties 972
disk-like molecule 942
disordered materials 38
dispersion
– glasses 543
dispersion curves
– electronic structure of surfaces 999
dispersion hardening 329
displacement of atoms
– surface phonons 1012
displacive disorder 38
displacive ferroelectrics 907
display lifetime 944
dissipation factor 826, 828
dissipative dispersion 907
dissociation energy of molecule
– elements 46
DOBAMBC (liquid crystal) 934
domain pattern
– perpendicular magnetization 1062
domain wall
– ferroelectric 907
– ferromagnetic 907
donor surface state 1022
doping
– chemical 576
DOS (density of states) 330, 1049
– nanostructured materials 1034
drain 1024
Drude-model metal 1045
drug delivery 944
DSP family 931
DTA (differential thermal analysis) 39
dubnium Db
– elements 105
ductile iron 269
Duran
– glasses 527, 531, 537
Dy dysprosium 142
dye dopant 944
dynamic viscosity 945–948, 955–959, 961, 962, 967, 969–971, 973
dysprosium Dy
– elements 142

E

E7
– liquid crystals 975
ECS 1011
EDX (energy-dispersive analysis of X-rays) 39
EELS 1013
effect of solute elements conductivity of Cu 297
effective masses
– aluminium compounds 617
– boron compounds 607
– cadmium compounds 681
– group IV semiconductors and IV–IV compounds 593
– indium compounds 646
– mercury compounds 689

– oxides of Ca, Sr, and Ba 661
– zinc compounds 670
effective masses m_n and m_p
– gallium compounds 629
einsteinium Es
– elements 151
elastic compliance 828
elastic compliance tensor 934
– elements 47
elastic constant c_{ik} 823
– aluminium compounds 611
– beryllium compounds 652
– cadmium compounds 676
– gallium compounds 621
– group IV semiconductors and IV–IV compounds 580
– indium compounds 638
– magnesium compounds 656
– mercury compounds 687
– oxides of Ca, Sr, and Ba 660
– zinc compounds 665
elastic modulus 478, *see* elastic constant 580
– elements 46, 47
– polymers 478
elastic stiffness 828
– elements 47
elastic tensor 827
elastooptic coefficient 825
elastooptic constant 828
elastooptic tensor 827
electric strength 478
– polymers 478
electrical conductivity
– aluminium compounds 618
– boron compounds 608
– elements 46
– group IV semiconductors and IV–IV compounds 595
electrical conductivity *see also* electrical resistivity 659
electrical conductivity σ
– indium compounds 647
– magnesium compounds 659
– mercury compounds 690
– oxides of Ca, Sr, and Ba 663
electrical resistivity
– boron compounds 609
– elements 48
– gallium compounds 630
electrical steel 766
electroforming 288
electromagnetic and optical properties
– group IV semiconductors and IV–IV compounds 601
– III–V compounds 610, 619, 635, 650
– II–VI compounds 655, 660, 664, 672, 683, 691
electromagnetic concentration effect 1046
electromagnetic confinement
– nanostructured materials 1044
electromechanical coupling constant 912, 918
electron affinity
– elements 46
electron and hole mobilities
– aluminium compounds 618
electron density of states 1034
– cadmium compounds 680
electron diffraction 41
electron effective mass m_n 593, *see* effective mass 661
electron g-factor g_c
– gallium compounds 629
– indium compounds 646
electron microscope image
– magnetic tunneling junction 1057
electron microscopy 1049
electron mobility μ_n 663, *see* mobility μ 682
– elements 48
– gallium compounds 631
– indium compounds 648
– mercury compounds 690
– oxides of Ca, Sr, and Ba 663
electron transport phenomena
– nanostructured materials 1042
electron tunneling
– phonon-assisted 1043
electronegativity
– elements 46
electronic band gap
– elements 48
electronic conductivity σ
– zinc compounds 671
electronic configuration
– elements 46
electronic dispersion curves 1000
electronic ground state
– elements 46
electronic properties
– group IV semiconductors and IV–IV compounds 589–594
– III–V compounds 606, 614, 626, 643
– II–VI compounds 653, 657, 661, 668, 678, 688
electronic structure
– solid surfaces 996
electronic transport, general description
– aluminium compounds 617
– beryllium compounds 655
– boron compounds 608
– cadmium compounds 682
– gallium compounds 629
– group IV semiconductors and IV–IV compounds 595
– indium compounds 647
– magnesium compounds 659
– mercury compounds 689
– oxides of Ca, Sr, and Ba 663
– zinc compounds 670
electronic work function
– elements 48
electronic, electromagnetic, and optical properties
– elements 46
electron–phonon coupling 1040, 1041
electrooptic coefficients 826
electrooptic modulators
– nanostructured materials 1040
electrooptic tensor 827
electro-optical constants
– zinc compounds 675
electro-optical effect 944
electrostrictive constant 930
elemental semiconductor 1003
elements 45
– allotropic modifications 48
– atomic properties 46
– electromagnetic properties 48
– electronic properties 48
– high-pressure modifications 48
– ionic properties 46
– macroscopic properties 46
– materials data 46
– molecular properties 46
– optical properties 48
– ordered according to the Periodic table 52
– ordered by their atomic number 51
– ordered by their chemical symbol 50
– ordered by their name 49
elinvar alloys 786
– antiferromagnetic 789
elinvar-type alloys
– nonmagnetic 792
energy bands *see* band structure 590

energy diagram for a MIM tunnel junction 1053
energy dispersion curve 1000
energy equivalents 24
energy equivalents in different units 24
energy exchange time τ_{e-ph} 1041
energy gap
– aluminium compounds 616
– beryllium compounds 654
– cadmium compounds 681
– gallium compounds 628
– group IV semiconductors 592, 593
– indium compounds 644
– IV–IV compound semiconductors 592, 593
– magnesium compounds 659
– mercury compounds 688
– oxides of Ca, Sr, and Ba 661
– zinc compounds 669
energy gaps
– boron compounds 607
energy shifts in the luminescence peaks 1037
energy-storage cell 1043
engineering critical current density 741
enthalpies of phase transitions 946–973
enthalpy change
– elements 47
enthalpy of combustion 477
– polymers 477
enthalpy of fusion 477
– polymers 477
entropy of fusion 477
– polymers 477
Er erbium 142
erbium Er
– elements 142
Es einsteinium 151
Eu europium 142
europium Eu
– elements 142
EXAFS (extended X-ray atomic fine-structure analysis) 39
excess carrier density 1023
excitation energy
– nanostructured materials 1038
exciton binding energy
– gallium compounds 628
– zinc compounds 669
exciton Bohr radii
– semiconductors 1037
exciton energy
– group IV semiconductors and IV–IV compounds 593
exciton peak energy
– indium compounds 645
exciton Rydberg series 1038
excitons
– nanostructured materials 1036
external forces 46
external-field dependence 46
extinction coefficient *k*
– gallium compounds 635
– zinc compounds 674

F

F fluorine 118
F2
– optical glasses 551
F5
– glasses 537
facets
– crystallography 27
Fahrenheit 48
families of ferroelectrics 909
fast displays 903
fcc positions
– surface diagrams 981
Fe iron 131
$Fe_{14}Nd_2B$
– commercial magnets 804
– magnetic materials 803
Fe–C(-X)
– carbide phases 224
Fe–Cr alloy 226
Fe–Mn alloys 226
Fe–Ni alloys 225
Fermi energy 998
Fermi function 1022
Fermi level pinning 1025
Fermi surface
– nanostructured materials 1055
Fermi surface shift in an electric field 1050
Fermi surfaces
– nanostructured materials 1049
Fermi wavelength
– nanostructured materials 1035
fermium Fm
– elements 151
ferrielectric triple hysteresis loop 936
ferrielectricity 927
ferrite
– applications 812
– hard magnetic 813
– MnZn 812
– NiZn 813
– soft magnetic 811
ferroelectric ceramics 906
ferroelectric hysteresis loop 904
ferroelectric liquid crystal 906, 911, 945, 967
– physical properties 967
ferroelectric mixtures
– liquid crystals 975
– physical properties 975
ferroelectric phase transition 906, 933
ferroelectric polymers 906
ferroelectric transducer 906
ferroelectrics
– classification 906
– definition 903
– dielectric properties 903
– displacive type 906
– elastic properties 903
– families 909, 911
– general properties 906
– indirect type 906
– inorganic crystals 903
– inorganic crystals other than oxides 922
– inorganic crystals oxides 912
– liquid crystals 903, 930
– order–disorder type 906
– organic crystals 903, 930
– phase transitions 903
– piezoelectric properties 903
– polymers 903, 930
– pyroelectric properties 903
– symbols and units 912
ferromagnetic surface 1008
Fe–Co–Cr 795
Fe–Co–V 797
Fe–Nd–B 800
– phase relations 800
Fe–Ni–Al–Co 798
Fe–Si alloys
– rapidly solidified 768
Fibonacci sequence 35
field-effect mobility 1024
– solid surfaces 1024
first Brillouin zone *see* Brillouin zone 657
flat glasses 528
flat-band condition 1020
fluorinated three-ring LC 944
fluorine F
– elements 118

fluoropolymers 480, 496
– poly(ethylene-co-chlorotrifluoroethylene) (ECTFE) 496, 497
– poly(ethylene-co-tetrafluoroethylene) (ETFE) 496, 497
– poly(tetrafluoroethylene-co-hexafluoropropylene) (FEP) 496, 497
– polychlorotrifluoroethylene (PCTFE) 496, 497
– polytetrafluoroethylene (PTFE) 496, 497
flux flow (FF) 718
Fm fermium 151
formation curve
– glasses 524
formulas
– crystallographic 986
Fotoceram
– glasses 559
Fotoform
– glasses 559
Foturan
– glasses 559
Fourier map 926
four-ring system 964
Fr francium 59
francium Fr
– elements 59
Frantz–Keldysh effect 1040
free dielectric constant 907
frequency conversion 825
Fresnel reflectivity
– glasses 549
Friedel–Creagh–Kmetz rule 943
fundamental constants 3
– 2002 adjustment 4
– alpha particle 9
– atomic physics and particle physics 7
– CODATA recommended values 4
– electromagnetic constants 6
– electron 7
– meaning 4
– most frequently used 4
– neutron 8
– proton 8
– recommended values 3, 4
– thermodynamic constants 6
– units of measurement 3
– universal constants 5
– what are the fundamental constants? 3
fused silica
– glasses 534, 537

G

Ga gallium 78
GaAs positions
– surface diagrams 983
gadolinium Gd
– elements 142
gallium antimonide
– crystal structure, mechanical and thermal properties 621
– electromagnetic and optical properties 635
– electronic properties 626
– transport properties 631
gallium arsenide
– crystal structure, mechanical and thermal properties 621
– electromagnetic and optical properties 635
– electronic properties 626
– transport properties 631
gallium compounds
– crystal structure, mechanical and thermal properties 621
– electromagnetic and optical properties 635
– electronic properties 626
– mechanical and thermal properties 621
– thermal conductivity 634
– thermal properties 621
– transport properties 629
gallium Ga
– elements 78
gallium nitride
– crystal structure, mechanical and thermal properties 621
– electromagnetic and optical properties 635
– electronic properties 626
– transport properties 631
gallium phosphide
– crystal structure, mechanical and thermal properties 621
– electromagnetic and optical properties 635
– electronic properties 626
– transport properties 631
gamma titanium aluminides 213
gas permeation 478
– polymers 478
Gd gadolinium 142
Ge germanium 88
germanium
– band structure 590
– crystal structure, mechanical and thermal properties 578–588
– electromagnetic and optical properties 601
– electronic properties 589–594
– transport properties 598
germanium Ge
– elements 88
g-factor
– cadmium compounds 681
g-factor, conduction electrons
– group IV semiconductors and IV–IV compounds 594
glass designation 544
glass formers 527
glass matrix 1040
glass number 8nnn 534, 537
glass number nnnn
– sealing glasses 563
glass structure
– sodium silicate glasses 524
glass temperature
– glasses 524
glass transition temperature 477
– polymers 477
glass-ceramics 525, 526, 558
– density 558
– elastic properties 558
– manufacturing process 558
glasses 523
– Abbe value 547
– abbreviating glass code 543
– acid attack 532
– acid classes 533
– alkali attack 532
– alkali classes 533
– alkali–alkaline-earth silicate 530
– alkali–lead silicate 530
– alkaline-earth aluminosilicate 530
– amorphous metals 523
– armor plate glasses 529
– automotive applications 529
– band pass filters 566
– Borofloat 528, 529
– borosilicate 529, 530
– borosilicate glasses 529
– brittleness 536
– chemical constants 553
– chemical properties 549
– chemical resistance 549
– chemical stability 530, 531
– chemical vapor deposition 523
– color code 554
– composition 527

– compound glasses 529
– container glasses 528, 529
– crack effects 536
– density 528
– dielectric properties 538
– Duran® 531
– elasticity 536
– electrical properties 537
– engineering material 523
– fire protecting glasses 529
– flat 528
– fracture toughness 537
– frozen-in melt 533
– halide glasses 568
– hydrolytic classes 533
– infrared transmitting glasses 568
– infrared-transmitting 571
– inhomogeneous 525
– internal transmission 554
– linear thermal expansion 536, 556
– long pass filters 566
– major groups 526
– manufacturers, preferred optical glasses 545
– melting range 533
– mixtures of oxide compounds 524
– neutral density filters 566
– optical 551
– optical characterization 547
– optical glasses 543
– optical properties 539, 543
– oxide glasses 568
– passivation glasses 562
– physical constants 553
– plate glasses 528
– properties 526
– quasi-solid melt 533
– refractive index 539
– Schott filter glasses 569
– sealing glasses 559
– short pass filters 566
– silicate based 526
– soda–lime type 528
– solder glasses 562
– strength 534
– stress behavior 535
– stress rate 535
– stress-induced birefringence 539
– supercooled melt 533
– surface (cleaning and etching) 533
– surface modification 532
– surface resistivity 538
– technical 530
– tensile strength 536
– thermal conductivity 556
– thermal strength 537
– transmittance 539
– viscosity 534
– vitreous silica 556
– volume resistivity 537
– wear-induced surface defects 535
glasses, colored
– nomenclature 566
– optical filter 566
glasses, sealing
– ceramic 562
– principal applications 561
– recommended material combinations 560
– special properties 561
glasses, solder and passivation
– composite 563
– properties 564
glassy state
– crystallography 39
GMO family
– ferroelectrics 920
GMR (giant magnetoresistance) 1049, 1050
– mechanism 1051
– thickness dependence 1051
GMR ratio 1050
gold
– alloys 347
– applications 347
– chemical properties 361
– electrical properties 356
– electrical resistivity 356
– intermetallic compounds 350
– magnetic properties 358
– mechanical properties 352
– optical properties 359
– phase diagrams 347
– production 347
– special alloys 361
– thermal properties 359
– thermochemical data 347
– thermoelectric properties 358
gold Au
– elements 65
golden mean 35
granular materials 1043
gray
– SI unit of absorbed dose 19
gray iron 269
gray tin
– band structure 590
– crystal structure, mechanical and thermal properties 578–588
– electromagnetic and optical properties 601
– electronic properties 589–594
– transport properties 599
Griffith flaw
– glasses 534
Group IV semiconductors 576, 578–603
– electron mobility 597
– hole mobility 597
Group IV semiconductors and IV–IV compounds
– crystal structure 578
– electromagnetic and optical properties 601
– electronic properties 589–594
– mechanical properties 578
– thermal properties 578
– transport properties 595–601
groups of elements (Periodic table) 45

H

H hydrogen 54
hafnium Hf
– elements 94
Haigh Push-Pull test 408
Hall coefficient
– elements 48
– group IV semiconductors and IV–IV compounds 595
Hall mobility
– group IV semiconductors and IV–IV compounds 596, 597
Hall mobility *see also* mobility μ 596
halogen-substituted benzene 946
hard disk drive 1060
– limits 1060
– technology 1060
hard ferrites
– magnetic properties 814
hard magnetic alloys 794
hardenability 237
hardmetals 277
hassium Hs
– elements 131
Hatfield steel 226
HATOF 1013
HCl family 924
He helium 54
heat capacities c_p, c_V
– boron compounds 605
– group IV semiconductors 584

– IV–IV compound semiconductors 584
heat capacity 477, 956–960, 970
– cadmium compounds 678
– gallium compounds 623
– indium compounds 640
– mercury compounds 687
– oxides of Ca, Sr, and Ba 661
– polymers 477
– zinc compounds 667
heat-resistant steels 258
HEIS (high-energy ion scattering) 989, 1013
helical structure 942
helium He
– elements 54
Hermann–Mauguin symbols 30
hertz
– SI unit of frequency 19
heterocycles 946
hexagonal $BaTiO_3$ 917
Hf hafnium 94
Hg mercury 73
high copper alloy 297
high-T_c superconductors
– lower critical 716
– upper critical 716
high-frequency dielectric constant ε *see also* dielectric constant ε 601
high-Ni alloys 281
high-pressure die casting (HPDC) 168
high-pressure modifications 48
high-strength low-alloy 240
hip implants 277
HMF (half-metallic ferromagnet) 1055
Ho holmium 142
hole effective mass m_p 594, *see* effective mass 661
hole mobility μ_p *see* mobility μ 682
– elements 48
– gallium compounds 631
– indium compounds 648
hollow-ware
– glasses 528
holmium Ho
– elements 142
holohedry
– crystallography 30
homeotropic alignment 943
Hooke's law 47, 823
hopping mechanism
– nanostructured materials 1043
host–guest effect 944
hot forming
– glasses 524
Hoya code
– glasses 544
HPDC (high-pressure die casting) 168
HRTEM (high-resolution transition electron microscopy) 39
Hs hassium 131
Hume-Rothery phase 333, 350
Hume-Rothery phases 296
hydrogen H
– elements 54
hydrostatic pressure 926
hyper-Raman scattering 915

I

I iodine 118
IBA 991, 1024
ICSU (International Council of the Scientific Unions) 4
ideal surface 979
III–V compound semiconductors 604
II–VI semiconductor compounds 652
image potential 996
image state 996, 997, 999
– effective mass 999
impact strength 478
– polymers 478
impurity elements 206
impurity scattering
– group IV semiconductors and IV–IV compounds 599
In indium 78
incommensurate phase 930
incommensurate phases 906
incommensurate reconstruction 986
index of refraction
– complex 674
indirect gap
– group IV semiconductors and IV–IV compounds 589
indium antimonide
– crystal structure, mechanical and thermal properties 638
– electromagnetic and optical properties 650
– electronic properties 643
– transport properties 647
indium arsenide
– crystal structure, mechanical and thermal properties 638
– electromagnetic and optical properties 650
– electronic properties 643
– transport properties 647
indium compounds
– crystal structure, mechanical and thermal properties 638
– electromagnetic and optical properties 650
– electronic properties 643
– mechanical and thermal properties 638
– optical properties 650
– thermal properties 638
– transport properties 647
indium In
– elements 78
indium nitride
– crystal structure, mechanical and thermal properties 638
– electromagnetic and optical properties 650
– electronic properties 643
– transport properties 647
indium phosphide
– crystal structure, mechanical and thermal properties 638
– electromagnetic and optical properties 650
– electronic properties 643
– transport properties 647
induced phase transition 904
inelastic neutron scattering 906
infrared-transmitting glasses 568, 571
inorganic ferroelectrics 903
inorganic ferroelectrics other than oxides 922
inosilicates 433
insulator
– surface phonon 1017
intercritical annealing 240
interface state density
– solid surface 1026
interlayer distance
– crystallographic formulas 986
International Annealed Copper Standard (IACS) 296
international system of units 11
international tables for crystallography 31
International Union of Pure and Applied Chemistry (IUPAC) 15
International Union of Pure and Applied Physics (IUPAP) 14
internuclear distance
– elements 46
intrinsic carrier concentration

– group IV semiconductors and IV–IV compounds 595, 596
intrinsic charge carrier concentration
– elements 48
intrinsic Debye length 1021
intrinsic Fermi level 1020
invar alloys
– Fe–Ni-based 782
– Fe–Pd base 785
– Fe–Pt-based 784
Invar effect 385
inversion center 30
inversion layer 1020
– solid surfaces 1024
inversion layer channel 1024
iodine I
– elements 118
ionic radius
– elements 46, 49
ionization energy
– elements 46
Ir iridium 135
iridium 393
– alloys 393
– applications 393
– chemical properties 398
– diffusion 398
– electrical properties 397
– lattice parameter 394
– magnetic properties 397
– mechanical properties 395
– optical properties 398
– phase diagram 393
– production 393
– thermal properties 398
– thermoelectrical properties 397
iridium Ir
– elements 135
iron and steels 221
iron Fe
– elements 131
iron miscibility gap 226
iron phase diagram 226
iron-carbon alloys 222
iron-cobalt alloys 772
iron–silicon alloys 763
Ising model 906
ISO (International Organization for Standardization) 12
isothermal transformation (IT) diagram 238
isotropic
– dielectrics 828
isotropic liquid 943
IV–IV compound semiconductors 576, 578–603
– electron mobility 597
– hole mobility 597

J

JDOS 1006
jellium 997
jellium model 998
– work functions 999
jewellery 330
joint density of states 1005
jominy apparatus 238
Josephson vortices 717
joule
– SI unit of energy 19

K

K 50
– glasses 537
K potassium 59
K10
– optical glasses 551
K7
– optical glasses 551
katal
– SI unit of catalytic activity 19
KDP family 925
kelvin 48
– SI base unit 14
Kerr effect 825
– optical 826
Kerr ellipticity 1011
KH_2PO_4 family 925
kilogram
– SI base unit 14
kinematic viscosity 945, 947, 950, 951, 955, 962, 964–967, 969, 975
Kleinman symmetry conditions 825
$KNbO_3$ 913
knee joint replacements 277
KNO_3 family 925
Knoop hardness
– optical glasses 550
Kr krypton 128
KRIPES 997
Kroll process 206
krypton Kr
– elements 128
$KTaO_3$ 913

L

LA
– phonon spectra 915
La lanthanum 84
Lamb theory of elastic vibrations 1040
lamellar (flake) graphite (FG) 268
langbeinite-type family 911, 930
lanthanum La
– elements 84
LASF35
– optical glasses 551
LAT family 932
lattice concept
– crystallography 28
lattice constants *see* lattice parameters 656
lattice dynamics 1012
lattice parameter 980
– aluminium compounds 610
– beryllium compounds 652
– boron compounds 604
– cadmium compounds 676
– gallium compounds 621
– group IV semiconductors and IV–IV compounds 579
– indium compounds 638
– magnesium compounds 656
– mercury compounds 686
– oxides of Ca, Sr, and Ba 660
– zinc compounds 665
lattice scattering
– group IV semiconductors and IV–IV compounds 598
lattice vibration 906
lattices
– planes and directions 28
Laue images 41
lawrencium Lr
– elements 151
layer-structure family 909
LB (Langmuir–Blodgett) film 944
LC display 944
LC materials (LCMs) 944
LC–surface interaction 944
lead 407
– antimony 412
– arsenic alloys 421
– battery grid alloys 413
– bearing alloys 415
– bismuth 419
– cable sheathing alloys 421
– calcium–tin 417
– calcium–tin, battery grid 418
– coper alloys 421
– corrosion 408
– corrosion classification 411
– fusible alloys 420
– gamma-ray mass-absorption 412
– grades 407

Subject Index

– internal friction 408
– low-melting alloys 419
– mechanical properties 408
– quaternary eutectic alloy 420
– recrystallization 409
– silver alloys 420
– solder alloys 415
– solders 416
– tellurium alloys 421
– ternary alloys 413
– tin alloy 415
lead glasses 527
lead Pb
– elements 88
lead–antimony
– phase diagram 413
LEED (low-energy electron diffraction) 987, 1013
LEIS (low-energy ion scattering) 989, 1013
LF5
– optical glasses 551
Li lithium 59
$Li_2Ge_7O_{15}$ family 920
$LiBaO_3$ 919
light transmittance
– glasses 539
light-emitting diode 1043
$LiNbO_3$ family 909, 919
linear thermal expansion
– optical glasses 556
linear thermal expansion coefficient α
– aluminium compounds 611
– beryllium compounds 653
– boron compounds 605
– cadmium compounds 677
– gallium compounds 623
– group IV semiconductors 582, 583
– indium compounds 639
– IV–IV compound semiconductors 582, 583
– magnesium compounds 656
– mercury compounds 687
– oxides of Ca, Sr, and Ba 660
– zinc compounds 666
$LiNH_4C_4H_4O_8$ family 911
liquid crystal
– anisotropy 943
– refractive index 943
– rod-like 943
liquid crystal family 911, 934
liquid crystal ferroelectrics 903
liquid crystal material (LCM)
– degradation 944
liquid crystal salts 973
– physical properties 973
liquid crystal two-ring systems with bridges
– physical properties 955
liquid crystal two-ring systems without bridges
– physical properties 947
liquid crystalline acids
– physical properties 946
liquid crystalline compound 943
– mesogenic group 943
– side group 943
– terminal group 943
liquid crystalline mixtures
– physical properties 975
liquid crystals
– ferroelectric properties 905
liquid crystals (LCs) 941
list of described physical properties 822
lithium Li
– elements 59
lithium niobate 919
lithography
– nanostructured materials 1064
Lithosil^{TM} 551
LLF1
– optical glasses 551
local-field effect 1045
long pass filters
– glasses 566
longitudinal acoustic branch 915
long-range order 39, 927
– glasses 524
long-range order of molecules 943
losses
– dynamic eddy current 763
– hysteresis 763
low dielectric loss glasses 527
low-dimensional system
– nanostructured materials 1034
low-frequency dielectric constant 917
low-temperature annealing 298
Lr lawrencium 151
LSMO ($La_{0.7}Sr_{0.3}MnO_3$) 1056
Lu lutetium 142
lumen
– non-SI unit in photometry 15
luminescence
– nanostructured materials 1036
luminous flux
– photometry 15
luminous intensity I_v
– photometry 15
lutetium Lu
– elements 142
lyotropic liquid crystals 942

M

M. Faraday 1032
Macor
– glasses 559
magnesium alloys 163
– corrosion behavior 169
– heat treatments 169
– joining 169
– mechanical properties 168
– nominal composition 165
– solubility data 163
– tensile properties 167
– tensile property 166
magnesium compounds
– crystal structure, mechanical and thermal properties 655
– electromagnetic and optical properties 660
– electronic properties 657
– mechanical and thermal properties 655
– optical properties 660
– thermal properties 655
– transport properties 659
magnesium Mg
– casting practices 168
– elements 68
– magnesium alloys 162
– melting practices 168
magnesium oxide 444
– applications 444
– crystal structure, mechanical and thermal properties 655
– electrical properties 444
– electromagnetic and optical properties 660
– electronic properties 657
– mechanical properties 444
– thermal properties 444
– transport properties 659
magnesium selenide
– crystal structure, mechanical and thermal properties 655
– electromagnetic and optical properties 660
– electronic properties 657
– transport properties 659
magnesium silicate
– electrical properties 435, 436
– mechanical properties 435, 436
– thermal properties 435, 436
magnesium sulfide

– crystal structure, mechanical and thermal properties 655
– electromagnetic and optical properties 660
– electronic properties 657
– transport properties 659
magnesium telluride
– crystal structure, mechanical and thermal properties 655
– electromagnetic and optical properties 660
– electronic properties 657
– transport properties 659
magnet
– Mn–Al–C 811
magnetic domain 1058
magnetic dots
– nanostructured materials 1049
magnetic dots, arrays of 1061
magnetic field constant
– fundamental constant 14
magnetic layers 1048, 1050
– spin valve 1052
magnetic materials 755
– Co_5Sm based 806
– hard 794
– permanent 794
magnetic nanostructures 1031, 1048, 1050
– information storage 1048
– read heads 1048
– sensors 1048
magnetic oxides 811
magnetic periodic structures 33
magnetic reading head 1053, 1058
magnetic recording
– perpendicular discontinuous media 1060
magnetic sensors 1048, 1058
magnetic surface 1008
magnetic susceptibility 948–950, 956, 957
– elements 46, 48
magnetic tunnel junction 1054
– manganite-based 1057
magnetic tunneling junctions (MTJ) 1056
magnetization
– elements 48
magnetocrystalline anisotropy 756
magnetoelectronic devices 1058
magnetoresistance effect 1050
magnetoresistance of Fe/Cr multilayers 1051
magnetostriction 757
magnets
– 17/2 type 807
– 5/1 type 806
– $TM_{17}Sm_2$ 808
majority carriers 1023
malleable irons 270
manganese Mn
– elements 124
manipulating the dot magnetization 1062
manufacturing
– Fe–Nd–B magnets 801
manufacturing process
– glasses 525
martensite 223
martensitic transformation 225, 226
mass magnetic susceptibility
– elements 48
mass susceptibility
– elements 48
mass-production glasses 526
materials
– semiconductors 576
materials data
– elements 46
material-specific parameters 46
matrix composites 170
Maxwell–Garnett model 1045
Maya blue color 1032
MBBA
– liquid crystals 955
MBE (molecular-beam epitaxy) 991, 1032, 1049, 1063
– 0-D structures 1064
– 1-D structures 1064
– 2-D structures 1063
Md mendelevium 151
mechanical and thermal properties
– III–V compounds 621, 638
– II–VI compounds 652, 655, 660, 665, 676, 686
mechanical properties
– elements 46, 47
– group IV semiconductors 578–588
– III–V compounds 610
– III–V semiconductors 604
– IV–IV compound semiconductors 578–588
– optical glasses 550
– technical glasses 533
MEIS (medium-energy ion scattering) 989, 1013
meitnerium Mt
– elements 135
melt viscosity 478
– polymers 478
melting point T_m
– aluminium compounds 611
– beryllium compounds 653
– boron compounds 605
– cadmium compounds 677
– gallium compounds 622
– group IV semiconductors and IV–IV compounds 580
– indium compounds 639
– magnesium compounds 656
– mercury compounds 686
– oxides of Ca, Sr, and Ba 660
– zinc compounds 666
melting temperature 477
– elements 47, 48
– polymers 477
memory devices 903
mendelevium Md
– elements 151
mercury compounds
– crystal structure, mechanical and thermal properties 686
– electromagnetic and optical properties 691
– electronic properties 688
– mechanical and thermal properties 686
– thermal properties 686
– transport properties 689
mercury Hg
– elements 73
mercury oxide
– crystal structure, mechanical and thermal properties 686
– electromagnetic and optical properties 691
– electronic properties 688
– transport properties 689
mercury selenide
– crystal structure, mechanical and thermal properties 686
– electromagnetic and optical properties 691
– electronic properties 688
– transport properties 689
mercury sulfide
– crystal structure, mechanical and thermal properties 686
– electromagnetic and optical properties 691
– electronic properties 688
– transport properties 689
mercury telluride
– crystal structure, mechanical and thermal properties 686

– electromagnetic and optical properties 691
– electronic properties 688
– transport properties 689
mesogen 943
mesogenic group 943
mesophase 941
mesoscopic material
– conductivity 1044
– nanoparticle doped 1044
– waveguide applications 1045
mesoscopic materials 1031, 1033
– manufacturing 1033
mesoscopic system
– quantum size effect 1035
– thermodynamic stability 1035
metal
– nanoparticle 1040
– resonance state 999
– surface 987
– surface core level shifts (SCLS) 998
– surface Debye temperature 1013
– surface phonon 1012
– surface state 999
– work function 997
metal surface 987
– jellium model 998
metals 997
– vertical relaxation 989
meter
– SI base unit 13
metrologica, international journal 12
MFM image of a written line on an array of dots 1063
MFM image of arrays of dots 1063
MFM image of domain pattern 1062
– sidewalls 1062
Mg magnesium 68
MHPOBC (liquid crystal) 935, 967
microphase separation 941
Mie theory 1045
Miller delta 825
Miller indices 28
M–I–M (metal–insulator–metal) heterostructure 1053
minority carriers 1023
missing-row reconstruction 987
Mn manganese 124
Mn−Al−C 810
– phase relations 810
Mo molybdenum 114
Mo-based alloys 317
mobility μ
– aluminium compounds 618
– cadmium compounds 682
– group IV semiconductors and IV–IV compounds 597–599
– indium compounds 648
MOCVD (metal-organic chemical vapor deposition) 1064
modulated crystal structure 924
modulated structures
– crystallography 34
moduli *see* elastic constant 580
Mohs hardness 822, 826
– elements 47
molar enthalpy of sublimation
– elements 47
molar entropy
– elements 47
molar heat capacity
– elements 47
molar magnetic susceptibility
– elements 48
molar mass 945–972
– glasses 527
molar susceptibility
– elements 48
molar volume
– elements 47
mole
– definition 14
– SI base unit 14
mole fraction
– glasses 527
molecular architecture 477
molybdenum Mo
– elements 114
momentum-conservation rule 1036
monocrystalline material 47
monolithic alloys 170
monophilic liquid crystal 942
MOS devices 979
MOS field-effect transistor 1020
MOSFET
– electron and hole mobility 1025
– equilibrium condition 1024
– schematic drawing 1025
Mott–Wannier exciton 1037
MQW (multiple quantum well) 1038
MRAM (magnetic random access memories) 1058
MRAM cell
– schematic diagram 1059
Mt meitnerium 135
MTJ (magnetic tunnel junction) 1053, 1054
MTJ sensor
– operation principle 1059
multi-component alloys 219
multiphase (MP) alloys 276
multiple hysteresis loops 932
multiple quantum wells 1042
MVA-TFT (multidomain vertical-alignment thin-film transistor) 944

N

N nitrogen 98
N16B
– glasses 537
– sealing glasses 563
N-4
– liquid crystals 959
Na sodium 59
$NaNO_2$ family 924
nanocrystal 1038
nanoimprint lithography 1061
nanoimprinted single domain dots
– images 1061
nanoimprinting 1061
nanolithography 1031
nanomaterial 1032
nanometric multilayers 1049
nanoparticle 1031
– doped material 1045
– local-field 1045
nanopatterning 1061
nanophase materials 1032
nanoporous materials 1032, 1043
nanoscience 1032
nanostructured material 1031, 1032
– classification scheme 1034
– conductance 1043
– definition 1032
– electrical conductivity 1043
– manufacturing 1063
– preparation 1031, 1035
– zeolites 1065
nanostructures
– magnetic 1048
nanotechnology 1032
Nb niobium 105
N-BAF10
– optical glasses 551
N-BAF52
– optical glasses 551
N-BAK4
– optical glasses 551
N-BALF4
– optical glasses 551
N-BASF64
– optical glasses 551

Nb-based alloys 318
N-BK7
– glasses 548
– optical glasses 551
Nd neodynium 142
Nd–Fe–B
– physical properties 802
Ne neon 128
near field microscopy 1049
nematic mixtures
– liquid crystals 975
– physical properties 975
nematic phase
– liquid crystals 941
nematic–isotropic transition 945
nematic-phase director 943
Neoceram
– glasses 559
neodymium Nd
– elements 142
neon Ne
– elements 128
Neoparies
– glasses 559
neptunium Np
– elements 151
nesosilicate 433
neutral density filters
– glasses 566
neutron diffraction 41, 926
neutron scattering 915
neutron spectrometer scan 916
new rheocast process (NRC) 170
newton meter
– SI unit of moment of force 19
N-FK51
– glasses 548
– optical glasses 551
N-FK56
– optical glasses 551
$(NH_4)_2SO_4$ family 927
$(NH_4)_3H(SO_4)_2$ family 928
$(NH_4)HSO_4$ family 928
$(NH_4)LiSO_4$ family 928
Ni nickel 139
Ni superalloys 294
nickel
– alloys 279
– application 279
– carbides 285
– low-alloy 279
– mechanical properties 280
– plating 288
nickel Ni
– elements 139
nickel-based superalloys 284
nickel–iron alloys 769
nickel-silvers 300
NIMs (National Institutes for Metrology 4
niobium Nb
– elements 105
nitride
– electrical properties 467
– mechanical properties 467
– thermal properties 467
nitrides
– physical properties 468
nitrogen N
– elements 98
N-KF9
– optical glasses 551
N-KZFS2
– optical glasses 551
N-LAF2
– optical glasses 551
N-LAK33
– optical glasses 551
N-LASF31
– optical glasses 551
No nobelium 151
nobelium No
– elements 151
noble metals 329
– Ag 329
– alloys 329
– applications 330
– Au 329
– catalysts 330
– corrosion resistance 329
– hardness 329
– Ir 329
– optical reflectivity 330
– Os 329
– Pd 329
– Pt 329
– Rh 329
– Ru 329
– vapour pressure 330
NOL (nano-oxide layer) 1053
noncrystallographic diffraction symmetries 36
nondestructive testing 941
nonlinear field-dependent properties 46
nonlinear optical coefficients
– nanostructured materials 1039
nonlinear optical device 903, 925
nonlinear optical susceptibility 920
nonlinear susceptibility tensors 827
non-oxide ferroelectrics 905
non-SI units 11, 20
normalizing 224
Np neptunium 151
N-PK51
– optical glasses 551
N-PSK57
– optical glasses 551
N-SF1
– optical glasses 551
N-SF56
– optical glasses 551
N-SF6
– glasses 548
– optical glasses 551
N-SK16
– optical glasses 551
N-SSK2
– optical glasses 551
nuclear incoherent scattering 916
nuclear reactor 218

O

O oxygen 108
occupied electron shells
– elements 46
Ohara code
– glasses 544
one-dimensional liquid 943
one-electron potential 999
opal
– nanostructured materials 1032
opals 1048
OPO (optical parametric oscillation) 825
optical constants
– gallium compounds 635
– group IV semiconductors and IV–IV compounds 601–603
– indium compounds 651
optical constants *n* and *k*
– cadmium compounds 684
optical glasses 526, 543
– thermal properties 556
optical materials
– high-frequency properties 824
optical mode frequency 916
optical parametric oscillation (OPO) 825
optical parametric oscillator 920
optical phonon scattering
– group IV semiconductors and IV–IV compounds 598
optical phonon scattering *see also* phonon scattering 598
optical phonon softening 913
optical properties

– group IV semiconductors and IV–IV compounds 601
– III–V compounds 610, 619, 635, 650
– II–VI compounds 655, 660, 664, 672, 683, 691
optical second-harmonic generator 903
optical transparency range 826
optical visualization 941
optoelectronic devices
– nanostructured materials 1038
order parameter 946–952, 956–960, 962, 963
order parameter, S 943
order principle for mesogenic groups 946
organic ferroelectrics 903
orientational order 941
Os osmium 131
osmium Os
– alloys 402
– applications 402
– cathodes 404
– chemical properties 406
– electrical properties 404
– elements 131
– lattice parameter 402
– magnetic properties 405
– mechanical properties 404
– phase diagrams 402
– production 402
– thermal properties 405
– thermoelectric properties 404
outer-shell orbital radius
– elements 46
over-aging 201
oxidation states
– elements 46
oxide 437
– beryllium 447
– magnesium 444
– physical properties 438
oxide ceramics
– production of 444
oxide ferroelectrics 903, 905
oxide superconductors
– low-T_c oxide 711
oxides of Ca, Sr, and Ba
– crystal structure, mechanical and thermal properties 660
– electromagnetic and optical properties 664
– electronic properties 661
– mechanical and thermal properties 660
– thermal properties 660
– transport properties 663
oxygen O
– elements 108

P

P phosphorus 98
Pa protactinium 151
PAA
– liquid crystals 958
pair distribution function
– crystallography 40
palladium Pd
– alloys 364
– applications 364
– electrical properties 370
– elements 139
– lattice parameters 366
– magnetic properties 372
– mechanical properties 368
– phase diagrams 364
– production 364
– thermoelectric properties 370
parabolic band 1000
Parkes process 330
partially stabilized zirconia (PSZ)
– electrical properties 449
– mechanical properties 449
– thermal properties 449
particle in a box model 1035
particle intensity I_p
– radiometry 15
passivation glasses 562, 564
– glasses 565
– properties 564
pattern transfer by imprinting 1061
Pauling 46
Pb lead 88
$PbHPO_4$ family 927
$PbTiO_3$ 916
$PbZrO_3$ 918
Pb–Ca–Sn
– battery grid alloys 418
PCH-7
– liquid crystals 950
Pd palladium 139
pearlite 222
Pearson symbol
– elements 47
percolation density
– nanostructured materials 1033
periodes of elements (Periodic table) 45
Periodic table
– elements 45
Periodic table of the elements 53
permanent magnets
– Co–Sm 806
perovskite-type family 909, 911, 912
perovskite-type oxide 909
phase diagram
– Fe–C 222
– Fe–Cr 226
– Fe–Mn 225
– Fe–Ni 225
– Fe–Si 227
– Ti–Al 210
– Zr–Nb 218
– Zr–O 218
phase separation
– glasses 525
phase transition temperature 943
phase-matching angle 826
phase-matching condition 825
phonon confinement
– nanostructured materials 1042
phonon density of states
– gallium compounds 623, 625
– indium compounds 642
phonon dispersion
– surface phonons 1012
phonon dispersion curve
– aluminium compounds 612
– gallium compounds 623–625
– indium compounds 641
– surface phonons 1015–1026
phonon dispersion relation
– group IV semiconductors 585–587
– IV–IV compound semiconductors 585, 588
phonon energies
– nanostructured materials 1040
– solid surfaces 1013
phonon frequencies ν
– cadmium compounds 678
– gallium compounds 624
– group IV semiconductors 585
– indium compounds 640
– IV–IV compound semiconductors 585
– magnesium compounds 657
– mercury compounds 687
– oxides of Ca, Sr, and Ba 661
– zinc compounds 667
phonon frequencies ν *see also* phonon wavenumbers $\tilde{\nu}$ 605
phonon instability 921
phonon mode frequency 908
phonon scattering

– group IV semiconductors and IV–IV compounds 598
phonon wavenumbers $\tilde{\nu}$ *see also* phonon frequencies ν 605
– aluminium compounds 612
– cadmium compounds 678
– gallium compounds 624
– indium compounds 640
– magnesium compounds 657
– mercury compounds 687
– zinc compounds 667
phonon wavenumbers $\tilde{\nu}$/frequencies ν
– boron compounds 605
phonon–phonon coupling 907
phosphorus P
– elements 98
photochromic glasses
– nanostructured materials 1032
photoconductive crystal 922
photoelastic effect 824
photoemission spectra for Ag quantum wells 1036
photoluminescence spectra of CdSe 1039
photometry
– intensity measurements 15
photonic band-gap materials 1048
phyllosilicate 433
physical properties
– liquid crystals 943
physical quantities 12
– base 11, 12
– data 13
– definition 12
– derived 11, 12
– general tables 4
piezoelectric
– element 918
– material 917
– strain constant 918
– strain tensor 824
– tensor 827
piezoelectricity 824, 906
piezooptic coefficient 825
planar alignment 943
planar electromechanical coupling factor 918
Planck radiator 15
plasmon excitations
– nanostructured materials 1031
plasmon oscillation
– nanostructured materials 1032
plasmon peak
– metals 1045
plasmon resonance 1045
plastic crystal 941
platinum group metals (PGM) 363
– alloys 363
platinum Pt
– alloys 376
– applications 376
– catalysis 385
– chemical properties 385
– electrical properties 381
– elements 139
– magnetical properties 384
– mechanical properties 378
– optical properties 385
– phase diagrams 376
– production 376
– thermal properties 385
– thermoelectric properties 382
plutonium Pu
– elements 151
PLZT
– ceramic material 918
Pm promethium 142
Po polonium 108
Pockels effect 825
point groups
– crystallography 30
Poisson equation 1022
Poisson number
– elements 47
Poisson's ratio 478
– optical glasses 550
– polymers 478
polarization microscopy 943
polonium Po
– elements 108
poly(4-methylpentene-1) (PMP) 488, 489
poly(ethylene-co-acrylic acid) (EAA) 486, 487
poly-(ethylene-co-norbornene) 486, 487
poly(ethylene-co-vinyl acetate) (EVA) 486, 487
poly(vinyl chloride) 492, 495
– plastisized (60/40) (PVC-P2) 492, 495
– plastisized (75/25) (PVC-P1) 492, 495
– unplastisized (PVC-U) 492, 494–496
polyacetals 480, 497
– poly(oxymethylene) (POM-H) 497–500
– poly(oxymethylene-co-ethylene) (POM-R) 497, 498, 500
polyacrylics 480, 497
– Poly(methyl methacrylate) (PMMA) 497–499
polyamides 480, 501
– polyamide 11 (PA11) 501
– polyamide 12 (PA12) 501
– polyamide 6 (PA6) 501, 502
– polyamide 610 (PA610) 501, 502
– polyamide 66 (PA66) 501, 502
polybutene-1 (PB) 488, 489
polyesters 481, 503
– poly(butylene terephthalate) (PBT) 503–505
– poly(ethylene terephthalate) (PET) 503–505
– poly(phenylene ether) (PPE) 504–506
– polycarbonate (PC) 503, 504
polyether ketones 481, 508
– poly(ether ether ketone) (PEEK) 508, 509
polyethylene 483–486
– high density (HDPE) 483–486
– linear low density (LLDPE) 484–486
– low density (LDPE) 484–486
– medium density (MDPE) 484–486
– ultra high molecular weight (UHMWPE) 484–486
polyethylene ionomer (EIM) 486, 487
polyimides 481, 508
– poly(amide imide) (PAI) 508
– poly(ether imide) (PEI) 508, 509
– polyimide (PI) 508
polyisobutylene (PIB) 488, 489
polymer
– physical properties 483
polymer blend 515
– physical properties 477, 483
– poly(acrylonitrile-co-butadiene-co-acrylester) + polycarbonate (ASA + PC) 515–517
– poly(acrylonitrile-co-butadiene-co-styrene) + polyamide (ABS + PA) 515–517
– poly(acrylonitrile-co-butadiene-co-styrene) + polycarbonate (ABS + PC) 515–517
– poly(butylene terephthalate) + poly(acrylonitrile-co-butadiene-co-acrylester) (PBT + ASA) 517, 521
– poly(butylene terephthalate) + polystyrene (PBT + PS) 515, 519, 520

– poly(ethylene terephthalate) + polystyrene (PET + PS) 515, 519, 520
– poly(phenylene ether) + polyamide 66 (PPE + PA66) 517, 521
– poly(phenylene ether) + polystyrene (PPE + PS) 520, 521
– poly(styrene-co-butadiene) (PPE + SB) 517, 520, 521
– poly(vinyl chloride) + chlorinated polyethylene (PVC + PE-C) 515, 518
– poly(vinyl chloride) + poly(acrylonitrile-co-butadiene-co-acrylester) (PVC + ASA) 515, 518
– poly(vinyl chloride) + poly(vinyl chloride-co-acrylate) (PVC + VC/A) 515, 518
– polycarbonate + liquid crystal polymer (PC + LCP) 515, 519, 520
– polycarbonate + poly(butylene terephthalate) (PC + PBT) 515, 519, 520
– polycarbonate + poly(ethylene terephthalate) (PC + PET) 515, 519, 520
– polypropylene + ethylene/propylene/diene rubber (PP + EPDM) 515, 516
– polysulfone + poly(acrylonitrile-co-butadiene-co-styrene) (PSU + ABS) 517, 521
polymer family 911, 936
polymer ferroelectrics 903
polymer matrix 1040
polymers 477
– abbreviations 482
– Charpy impact strength 478
– coefficient of expansion 478
– compressibility 478
– creep modulus 478
– crystallinity 477
– density 478
– elastic modulus 478
– electric strength 478
– enthalpy of combustion 477
– enthalpy of fusion 477
– entropy of fusion 477
– ferroelectric properties 905
– gas permeation 478
– glass transition temperature 477
– heat capacity 477
– impact strength 478
– melt viscosity 478
– melting temperature 477
– physical properties 477
– physicochemical properties 477
– Poisson's ratio 478
– refractive index 478
– relative permittivity 478
– shear modulus 478
– shear rate 478
– Shore hardness 478
– sound velocity 478
– steam permeation 478
– stress 478
– stress at 50% strain (elongation) 478
– stress at fracture 478
– stress at yield 478
– structural units 479–481
– surface resistivity 478
– thermal conductivity 478
– Vicat softening temperature 477
– viscosity 478
– volume resistivity 478
polyolefines 480, 483–486
polypropylene (PP) 488, 489
polysulfides 481, 506
– poly(phenylene sulfide) (PPS) 506, 507
polysulfones 481, 506
– poly(ether sulfone) (PES) 506, 507
– polysulfone (PSU) 506, 507
polyurethanes 481, 511
– polyurethane (PUR) 511, 512
– thermoplastic polyurethane elastomer (TPU) 511, 512
polyvinylidene fluoride
– ferroelectrics 911
porous aluminium silicates
– electrical properties 436
– mechanical properties 436
– thermal properties 436
Portland cement
– ASTM types 433
– chemical composition 432
positional order 941
potassium K
– elements 59
potential barrier 1022
powder-composite materials 277
Powder-in-Tube (PIT) 739
power-law dependence of conductivity on film thickness 1042
Pr praseodynium 142
practical superconductors
– characteristic properties 705
praseodynium Pr
– elements 142
prefixes
– decimal multiples of units 19
primitive cell
– crystal structure 28
projected band structure 996
projected bond length 995
promethium Pm
– elements 142
property tensor 46
– independent components 827
protactinium Pa
– elements 151
proton distribution 926
pseudopotential calculation 1006
Pt platinum 139
p-type diamond
– Debye length 1021
Pu plutonium 151
pulsed infrared lasers 1040
pyrochlore-type family 909
pyroelectric coefficient 917
pyroelectric measurement 931
PZT
– piezoelectric material 917

Q

quantum confinement 1036
– nanostructured materials 1042
quantum dots
– nanostructured materials 1031, 1035
quantum size effect
– nanostructured materials 1031, 1035
quantum transport
– nanostructured materials 1053
quantum well
– coupled 1043
– nanostructured materials 1031, 1035
quantum wires
– nanostructured materials 1035
quantum-well superlattices 1033
quasicrystals 34
QWIP (quantum well infrared photodetector) 1042

R

Ra radium 68
radiant intensity I_e
– radiometry 15

radiation sources and exposure techniques in lithography 1065
radiometric and photometric quantities 16
radiometry
– intensity measurements 15
radium Ra
– elements 68
radon Rn
– elements 128
Raman scattering 906
Raman scattering spectroscopy 1040
Raman spectrum 908
RAS 997
Rayleigh mode 1012
Rb rubidium
– elements 59
Re rhenium
– elements 124
real and imaginary parts ε_1 and ε_2 of the dielectric constant *see* dielectric constant ε 602
– gallium compounds 637
reconstruction model 987
– solid surfaces 991
reconstruction of semiconductors 991
reconstruction of surface 986
– metals 987
recording media
– arrays of magnetic dots 1061
reduced surface state energy 1022
reduced wave vector 1012
reduced-dimensional material geometries 1033
references
– solid surfaces 1029
reflectance anisotropy spectroscopy (RAS) 1007
refractive index 478, 829, 946–972
– elements 48
– glasses 539, 543
– polymers 478
– Sellmeier dispersion formula 547
– temperature dependence 548
refractive index n 619
– boron compounds 610
– cadmium compounds 684
– gallium compounds 635
– group IV semiconductors and IV–IV compounds 601–603
– indium compounds 650
– mercury compounds 691
– zinc compounds 672
refractories
– boride-based 452
– carbide-based 458
– nitride-based 468
– oxide-based 438
– silicide-based 472
refractory ceramics 437
refractory metals 303
– alloys 303
  compositions 305
  dispersion-strengthened 304
– annealing 311
– chemical properties 308
– crack growth behavior 325
– creep elongation 316
– creep properties 327
– dynamic properties 318
– evaporation rate 307
– fatigue data 321
– flow stress 316
– fracture mechanics 322
– grain boundaries 314
– high-cycle fatigue properties 319
– linear thermal expansion 306
– low-cycle fatigue properties 320
– mechanical properties 314
– metal loss 308
– microplasticity 318
– oxidation behavior 308
– physical properties 306
– production routes 304
– recrystallization 311
– resistance against gaseous media 309
– resistance against metal melts 309
– specific electrical resistivity 307
– specific heat 307
– static mechanical properties 315
– stress–strain curves 320
– thermal conductivity 306
– thermomechanical treatment 314
– vapor pressure 307
– Young's modulus 307
refractory metals alloys
– application 306
– products 306
refractory production
– raw materials 444
relative permittivity 478
– polymers 478
relaxation of semiconductors 991
relaxation of surface 986
– metals 987
relaxor 906, 909, 918
remanent magnetization 922
remanent polarization 918
residual resistance ratio (RRR) 397
residual resistivity ratio (RRR) 338
resistivity
– gallium compounds 630
response of material 46
Rf rutherfordium 94
Rh rhodium
– elements 135
RHEED (reflection high-energy electron diffraction) 990
rhenium Re
– elements 124
rhodium
– alloys 386
– applications 386
– chemical properties 392
– electrical properties 390
– magnetical properties 391
– mechanical properties 387
– optical properties 392
– phase diagrams 386
– production 386
– thermal properties 392
– thermoelectrical properties 391
rhodium Rh
– elements 135
ribbon silicates 433
RIE (reactive ion etching) 1061
Rn radon 128
Rochelle salt 904
Rochelle salt family 932
rod-like molecule 942
RT (room temperature) 49
RTP (room temperaure and standard pressure) 49
Ru ruthenium 131
rubidium Rb
– elements 59
ruthenium Ru
– alloys 399
– applications 399
– chemical properties 402
– electrical properties 401
– elements 131
– lattice parameter 400
– magnetic properties 401
– mechanical properties 400
– optical properties 402
– phase diagrams 399
– production 399
– thermal properties 402
– thermoelectric properties 401
rutherfordium Rf
– elements 94
RW (weighted sound reduction) 409

S

S sulfur 108
SAE (Society of Automotive Engineers) 221
SAM (self-assembled monolayer) 944
samarium Sm
– elements 142
SAW (surface acoustic wave) 912
Sb antimony 98
SbSI family 922
Sc scandium 84
$SC(NH_2)_2$ family 930
scandium Sc
– elements 84
scattering
– nanoscale objects 1048
scattering losses of a waveguide 1045
Schoenflies symbol 30
– elements 47
Schott AG 523
Schott code
– glasses 544
Schott filter glasses
– glasses 569
Schott glasses 8nnn 540, 541
SDR 997
Se selenium 108
seaborgium Sg
– elements 114
sealing glasses 527
– glasses 559
second
– SI base unit 14
secondary hardening 263
second-harmonic generation (SHG) 825, 906
second-order elastic constants *see* elastic constant 580
second-order phase transition 907
selenium Se
– elements 108
Sellmeier dispersion formula
– glasses 547
Sellmeier equations 826
SEM (scanning electron microscopy) 39
semiconductor 1003
– covalent 992
– field-effect mobility 1024
– III–V compounds 1004
– intrinsic Debye length 1021
– nanocrystal 1040
– polar 992
– quantum confinement 1036
– reconstruction 1004
– reconstruction model 991
– surface 990
– surface core level shift 1003
– surface Debye temperature 1017
– surface phonon 1017
– surface shift 1004
semiconductor band bending 1020
semiconductor nanostructures 1035
semiconductor surface
– Fermi level pinning 1025
– ionization energy 1003
semiconductors
– aluminium compounds 610
– boron compounds 604
– cadmium compounds 676
– chemical doping 576
– gallium compounds 621
– group IV semiconductors and IV–IV compounds 578–603
– III–V compounds 576, 604
– II–VI compounds 576, 652
– indium compounds 638
– introduction 575
– IV–IV compounds 576
– magnesium compounds 655
– mercury compounds 686
– oxides of Ca, Sr, and Ba 660
– physical properties 577
– table of contents 575
– zinc compounds 665
semi-solid metal processing (SSMP) 170
sensor
– chemiresistor-type 1043
SF1
– optical glasses 551
SF11
– optical glasses 551
SF2
– optical glasses 551
SF6
– glasses 537, 548
– optical glasses 551
SF66
– optical glasses 551
SFG (sum frequency generation) 825
SFM (superfluorinated material) 975
Sg seaborgium 114
shape memory 298
– nickel 279
shape-memory alloys
– TiNi 216
shear modulus 478
– elements 47
– polymers 478
shear rate 478
– polymers 478
SHG (second-harmonic generation) 825, 906
Shore hardness 478
– polymers 478
short pass filters
– glasses 566
short-range order 39
– glasses 524
Shubnikov groups
– crystallography 33
SI (Système International d'Unités) 3, 11
SI (the International System of Units) 12
SI base unit 13
SI definitions of magnetic susceptibility 48
SI derived units 16, 17
– with special names 17, 18
SI prefixes 19
Si silicon 88
SI units
– base quantities 13
– base units 13
Si_3N_4 ceramics 451
Si_3N_4 powders 472
SiC ceramics 451
side group 943
sievert
– SI unit of dose equivalent 19
silica
– glasses 524
silicate 433
silicate based glasses 526
silicide 473
– physical properties 472
silicon
– electromagnetic and optical properties 601
– electronic properties 589–594
– transport properties 598
silicon carbide
– band structure 590
– crystal structure, mechanical and thermal properties 578–588
– electromagnetic and optical properties 601

– electronic properties 589–594
– transport properties 595
silicon nitride 467
silicon Si
– crystal structure, mechanical and thermal properties 578–588
– elements 88
silicon steels
– grain-oriented 765
– non-oriented 763
silicon technology 1036
silicon-based lasers 1036
silicon–germanium alloys
– band structure 590
– transport properties 601
silicon-germanium alloys
– crystal structure, mechanical and thermal properties 578–588
– electromagnetic and optical properties 601
silicon–silicon oxide interface 1025
silver 330
– alloys 330
– application 330
– chemical properties 344
– crystal structures 333
– diffusion 342
– electrical properties 338
– intermetallic phases 333
– magnetic properties 339
– mechanical properties 335
– optical properties 341
– phase diagrams 331
– production 330
– ternary alloys 345
– thermal properties 340
– thermodynamic data 331
– thermoelectric properties 339
silver Ag
– elements 65
simple perovskite-type oxide 909
single hysteresis loop 904
SiO_2
– glasses 524
SK51
– optical glasses 551
Sm samarium
– elements 142
smectic C* phase 934
smectic phase 942
Sn tin 88
SNR (signal-to-noise ratio) 1060
soda lime glasses 528, 529, 534
sodium Na
– elements 59
soft annealing 224
soft magnetic alloys 758
– nanocrystalline 776
soft magnetic materials
– composite 759
– sintered 759
soft-mode spectroscopy 906
solar cell 1043
solder alloy 345
solder glasses 562
sol–gel synthesis
– nanostructured materials 1065
solid material
– structure 27
solid material, structure 27
solid surface energy 944
solid-state polymorphism 943
sorosilicate 433
sound velocity 478, 947–949, 956–962, 968–971
– elements 47
– polymers 478
source 1024
sp^3-bonded crystal 991
space charge function
– solid surfaces 1022
space charge layer
– semiconductor surface 1020
– solid surfaces 1020
space groups
– crystallography 31
SPARPES 1011
speed of light
– fundamental constant 13
spheroidal (nodular) graphite (SG) 268
spheroidite 223
spin accumulation 1049
spin diffusion length 1050
spin electronics
– applications 1057
– nanostructured materials 1031, 1049
spin polarization 1055
– nanostructured materials 1049
spin valve multilayers 1052
spin valve read head
– schematic diagram 1058
spin valve sensor 1053
spin-asymmetric material 1049
spinel structure
– crystallography 33
spin-electronic switch 1055
spin–orbit splitting energy Δ_{so}
– aluminium compounds 617
– cadmium compounds 681
– gallium compounds 628
– group IV semiconductors and IV–IV compounds 593
– indium compounds 644
– mercury compounds 689
– zinc compounds 670
spintronics
– nanostructured materials 1031, 1049
SPLEED 1011
spontaneous electric polarization 903
spontaneous polarization 917, 945
Sr strontium 68
$Sr_2Nb_2O_7$ family 920
SRI (sound reduction index) 409
$SrNb_2O_7$ family 909
$SrTeO_3$ family 919
$SrTiO_3$ 913
stabilized zirconia (PSZ) 448
stacking faults
– crystallography 41
stain resistance
– optical glasses 550
stainless steels 240
– austenitic 252
– duplex 257
– ferritic 246
– martensitic 250
– martensitic-ferritic 250
standard electrode potential
– elements 46
standard entropy
– elements 47
standard temperature and pressure (STP) 46
Stark effect
– nanostructured materials 1040
static dielectric constant 826
– elements 48
static dielectric constant 828–889
STC (sound transmission classification) 409
steam permeation 478
– polymers 478
steel
– austenitic 259
– carbon 227
– ferritic 258
– ferritic austenitic 259
– hardening 237
– heat-resistant 258, 261
– high-strength low-alloy (HSLA) 240

– low-alloy carbon steel 227
– mechanical properties 237
– stainless 240
– tool 262
stibiotantalite family 909
stiffness constant *see* elastic constant 580
STM (scanning tunneling microscopy) 988, 992, 997
STM spectroscopy 1005
STN (supertwisted nematic) effect 944
STO ($SrTiO_3$) 1056
storage capacity of hard disks 1048
storage density evolution of hard disk drives 1048
storage media 1060
– arrays of nanometer-scale dots 1060
– limits 1060
– technology 1060
storing information on the sidewalls of the dots 1062
strain
– polymers 478
strength
– glasses 534
stress 478
– polymers 478
stress at 50% strain (elongation) 478
– polymers 478
stress at fracture 478
stress at yield 478
– polymers 478
stress birefringence
– glasses 539, 549
stress intensity factor
– glasses 534
strong-confinement regime
– nanostructured materials 1037, 1038
strontium oxide
– crystal structure, mechanical and thermal properties 660
– electromagnetic and optical properties 664
– electronic properties 661
– transport properties 663
strontium Sr
– elements 68
strontium titanate 913
structural parameters 990
structural phase transitions 906
structure
– diamond-like 979
structure type
– crystallography 33
Strukturbericht type 47
sublattice 903
sublattice polarization 904
submicrometer magnetic dots 1060
substituted mesogens (liquid crystals)
– physical properties 968
sulfur S
– elements 108
sum frequency generation (SFG) 825
superalloys
– Ni-based cast 288
– nickel 294
superconducting high-T_c
– crystal structure 712
superconducting oxides
– high-T_c chemical composition 712
superconductivity
– elements 48
superconductor 695
– borides 745
– borocarbides 746, 747
– carbides 745
– commercial Nb_3Sn 709
– critical temperature 699
– crystal structure 712
– Debye temperature 696
– device applications 719
– high-T_c cuprates 712, 713, 720
– industrial wire performance 719
– metallic 696
– Nb alloys 702
– non-metallic 712
– Pb alloys 696
– pinning 717
– practical metallic 704
– production Nb_3Sn 708
– Sommerfeld constant 696
– SQUIDs 720
– structural data 720
– thermodynamic properties 696
– Type I 695
– Type II 695
– V alloys 700
– vortex lines 717
– Y–Ba–Cu–O 723
supercooled liquid
– glasses 524
supercooled mesophase 945
superstructures
– crystallography 41
supertwisted nematic (STN) effect 944
Supremax
– glasses 527
surface
– Curie temperature 1009
– diagram 979
– ionization energy 1003
– magnetic 1008
– semiconductor 990
– structure of an ideal 979
surface band structure 996
surface Brillouin zone (SBZ) 996
surface conductivity
– solid surfaces 1024
surface core level shifts (SCLS) 998
– solid surfaces 1003, 1004
surface differential reflectivity (SDR) 1008
surface excess conductivity 1024
surface magnetization 1011
surface mobility 1024
surface of diamond 1004
surface phonon 1012
– dispersion 1019
– metal 1013
– mode 1012
surface plasmon
– absorption of nanoparticles 1046
– dispersion curve 1001
surface resistivity 478
– polymers 478
surface resonance 996
– phonons 1012
surface response
– dielectric theory 1007
surface state
– acceptor 1022
– band 1005
– donor 1022
– transitions 1005
surface state bands
– solid surfaces 1004
surface states 996
surface tension 948–950, 955–962, 968–971, 975
– elements 47
surface tension (γ_{LV}) 943
surfaces 979
surgical implant alloys
– cobalt-based 277
susceptibility

– magnetic 48
– mass 48
– molar 48
– nonlinear dielectric 825, 829
– second-order nonlinear dielectric 826
– third-order nonlinear dielectric 826
SV (spin valve) 1052
symmetry elements of point groups 30
synthesis of clusters
– gas-phase production 1066
– nanostructured materials 1065
synthetic silica
– glasses 557

T

T tritium 54
TA
– phonon spectra 915
Ta tantalum 105
Ta-based alloys 318
tailoring of the electronic wave function 1032
tantalum Ta
– elements 105
Tb terbium 142
TC 12 (Technical Committee 12 of ISO) 12
Tc technetium 124
Te tellurium 108
technetium Tc
– elements 124
technical ceramics 437
technical coppers 297
technical glasses 527, 530
technical specialty glasses 526
tellurium Te
– elements 108
TEM (transmission electron microscopy) 39
TEM image of a superlattice of Au clusters 1066
TEM views of single dots 1062
temper graphite (TG) 268
temperature dependence of carrier concentration
– group IV semiconductors and IV–IV compounds 596
temperature dependence of electrical conductivity
– indium compounds 647
temperature dependence of electronic mobilities
– indium compounds 649
temperature dependence of energy gap
– indium compounds 645
temperature dependence of linear thermal expansion coefficient
– cadmium compounds 677
– magnesium compounds 656
temperature dependence of the lattice parameters
– group IV semiconductors and IV–IV compounds 580–582
temperature dependence of thermal conductivity
– group IV semiconductors and IV–IV compounds 599, 600
– indium compounds 650
temperatures of phase transitions 946–976
tempering of steel 223
template synthesis 944
tensile strength
– elements 47
tensor
– elastooptic 826
– piezoelectric strain 826
terbium Tb
– elements 142
terminal group 943
ternary alloys 298
terne steel coatings 415
TFT (thin-film transistor) 944
TGS family 932
Th thorium 151
thallium Tl
– elements 78
thermal and thermodynamic properties
– elements 46
thermal conductivity κ 478, 945, 947–949, 956, 958
– aluminium compounds 618
– cadmium compounds 682
– elements 46, 47
– gallium compounds 634
– group IV semiconductors and IV–IV compounds 599
– indium compounds 650
– magnesium compounds 659
– mercury compounds 690
– oxides of Ca, Sr, and Ba 663
– polymers 478
– zinc compounds 670
thermal expansion
– glasses 526
thermal expansion coefficient, linear
– elements 47
thermal gap
– indium compounds 645
thermal properties
– group IV semiconductors 578–588
– III–V compounds 610, 621, 638
– III–V semiconductors 604
– II–VI compounds 652, 655, 660, 665, 676, 686
– IV–IV compound semiconductors 578–588
– technical glasses 533, 536
thermal vibrations
– surface phonons 1012
thermal work function
– elements 48
thermally activated flux flow (TAFF) 718
thermochromic material 944
thermodynamic properties
– elements 47
thermoelectric coefficient
– elements 48
thermoelectric power
– oxides of Ca, Sr, and Ba 664
thermography 941, 944
thermomechanical treatment (TMT) 314
thermosets 481, 512
– diallyl phthalate (DAP) 512, 514
– epoxy resin (EP) 514, 515
– melamine formaldehyde (MF) 512, 513
– phenol formaldehyde (PF) 512, 513
– polymers 512
– silicone resin (SI) 514, 515
– unsaturated polyester (UP) 512–514
– urea formaldehyde (UF) 512, 513
thermotropic liquid crystal 942
thin film 906
thin-film transistor (TFT) 944
thixomolding 170
thorium Th
– elements 151
three and four-ring systems
– liquid crystals 964
three-dimensional long-range order 941
three-ring system 964
three-wave interactions
– in crystals 825
thulium Tm
– elements 142

Ti titanium 94
time-temperature-transformation (TTT) diagram 238
tin Sn
– elements 88
titanates 450
titanium 206
– commercially pure grades 207
– creep behavior 208
– creep strength 210
– hardness 207
– high-temperature phase 206
– intermetallic materials 210
– phase transformation 206
– sponge 207
– superalloys 210
– titanium alloys 206
titanium alloys 209
– applications 209
– chemical composition 209
– chemical properties 213
– mechancal properties 213
– mechanical properties 209
– physical properties 213
– polycrystalline 213
– single crystalline 213
– thermal expansion coefficient 214
titanium dioxide
– mechanical properties 450
– thermal properties 450
titanium oxide
– phase diagram 206
titanium Ti
– elements 94
Tl thallium 78
Tm thulium 142
TMR (tunnel magnetoresistance) 1054, 1055
TN (twisted nematic)
– liquid crystals 944
TO 915
tool steels 262
torsional modulus
– optical glasses 550
total losses 763
transformation temperature
– glasses 524
transition range
– glasses 524
transition temperature
– glasses 525
transitions
– surface states 1005
transmission spectra
– colored glasses 566, 567
transmission window
– glasses 524
transmittance
– glasses 548
transmittance of glasses
– color code 549
transport properties
– group IV semiconductors and IV–IV compounds 595–601
– III–V compounds 608, 617, 629, 647
– II–VI compounds 655, 659, 663, 670, 682, 689
transverse acoustic branch 915
transverse optical branch 915
transverse optical mode 906
triple point of water 48
tritium T
– elements 54
truncated crystal 986
tungsten bronze-type family 909, 920
tungsten W
– elements 114
tunnel junction
– magnetic 1053
tunnel magnetoresistance
– function of field and temperature 1057
tunnel magnetoresistance as a function of magnetic field 1055
tunneling
– nanostructured materials 1053
tunneling mechanism
– nanostructured materials 1043
twisted nematic (TN) effect 944
two-dimensional liquid 942
two-photon absorption coefficient 829
two-ring systems with bridges
– liquid crystals 955
two-ring systems without bridges
– liquid crystals 947
Type II superconductors
– anisotropy coefficients 716
– coherence lengths 716
– high-T_c cuprate compounds 716
type metals 414

U

U uranium 151
ultrahigh density storage media 1049, 1060
unalloyed coppers 296
uniaxial crystals 826
Unified Numbering System for Metals and Alloys (UNS) 296
unit cell of Si(111) 7×7 995
units
– amount of substance 14
– atomic 21
– atomic units (a.u.) 22
– candela 15
– CGS units 21
– coherent set of 20
– crystallography 21
– electric current 14
– general tables 4
– length 13
– luminous intensity 15
– mass 14
– natural 21
– natural units (n.u.) 21
– non-SI 22
– non-SI units 20, 21
– other non-SI units 23
– temperature 14
– the international system of 11
– time 14
– used with the SI 20
– X-ray-related units 22
units of physical quantities
– fundamental constants 3
units outside the SI 20
UNS (Unified Numbering System) 221
UPS 998
uranium U
– elements 151
UTS – ultimate tensile strength 219

V

V vanadium 105
van der Waals attraction 1019
vanadium V
– elements 105
vertical nanomagnets 1062
vertical relaxation of metals 989
VFT (Vogel, Fulcher, Tammann) equation
– glasses 533
Vicat softening temperature 477
– polymers 477
Vickers hardness
– elements 47

vinylpolymers 480, 489–492
– poly(acrylonitrile-co-butadiene-co-styrene) (ABS) 492–494
– poly(acrylonitrile-co-styrene-co-acrylester) (ASA) 492–494
– poly(styrene-co-acrylnitrile) (SAN) 489, 490, 492
– poly(styrene-co-butadiene) (SB) 489–491
– poly(vinyl carbazole) (PVK) 492, 493
– polystyrene (PS) 489–491
VIP (viewing-independent panel) 975
viscosity 478, 948, 950–954, 963–965, 967, 968, 975
– dynamic 943
– elements 47
– glasses 524
– kinematic 943
– optical glasses 556
– polymers 478
– technical glasses 533
– temperature dependence 525
viscosity of glasses
– temperature dependence 534
vitreous silica
– electrical properties 557
– gas solubility 557
– glasses 526, 556
– molecular diffusion 557
– optical constants 557
vitreous solder glasses 562
Vitronit
– glasses 559
volume compressibility
– elements 47
volume magnetization
– elements 48
volume of primitive cell
– crystallographic formulas 986
volume resistivity 478
– polymers 478
volume–temperature dependence
– glasses 524
Vycor™
– glasses 527

W

W tungsten 114
wavelength dependence of refractive index n
– indium compounds 651
WDX (wavelength-dispersive analysis of X-rays) 39
weak-confinement regime
– nanostructured materials 1037, 1038
wear-induced surface defects
– glasses 535
Weibull distribution
– glasses 535
weight fraction
– glasses 527
Wood's metal 420
work function Φ
– metal 997
– solid surfaces 997
work hardening wrought copper alloys 300
wrought alloys 298
wrought magnesium alloys 164
wrought superalloys 284
wtppm (weight part per million) 407
Wyckoff position
– crystallography 32

X

Xe xenon 128
xenon Xe
– elements 128
X-ray diffraction 39
X-ray interferences
– crystallography 27

Y

Y yttrium 84
Yb ytterbium 142
Young's modulus 823
– elements 47
– optical glasses 550
YS – yield stress 219
ytterbium Yb
– elements 142
yttrium Y
– elements 84
Y–Ba–Cu–O
– critical current density 733
– crystal defects 728
– crystal structure 723
– electric resistivity 730
– grain boundaries 730
– hole concentration 733
– lattice parameters 726
– lower critical field 734
– oxygen content 724
– pinning 735
– substitutions 725
– superconducting properties 731
– thermal conductivity 730
– transition temperature 733
– upper critical field 734

Z

zeolites
– nanostructured materials 1031, 1065
Zerodur®
– glasses 558
– linear thermal expansion 558
zinc compounds
– crystal structure, mechanical and thermal properties 665
– effective hole mass 670
– electromagnetic and optical properties 672
– electronic properties 668
– mechanical and thermal properties 665
– optical properties 672
– thermal properties 665
– transport properties 670
zinc oxide
– crystal structure, mechanical and thermal properties 665
– electromagnetic and optical properties 672
– electronic properties 667
– transport properties 670
zinc selenide
– crystal structure, mechanical and thermal properties 665
– electromagnetic and optical properties 672
– electronic properties 667
– transport properties 670
zinc sulfide
– crystal structure, mechanical and thermal properties 665
– electromagnetic and optical properties 672
– electronic properties 667
– transport properties 670
zinc telluride
– crystal structure, mechanical and thermal properties 665
– electromagnetic and optical properties 672
– electronic properties 667
– transport properties 670

zinc Zn
– elements 73
zircaloy 219
– irradiation effect 219
zirconium
– alloys 217
– bulk glassy alloys 218
– bulk glassy behavior 220
– low alloy materials 217
– nuclear applications 218
– technically-pure materials 217
zirconium dioxide 448
zirconium Zr
– elements 94
ZLI-1132
– liquid crystals 975
Zn zinc 73
Zr zirconium 94

Periodic Table of the Elements

18 VIIIA	IUPAC Notation CAS Notation
2 He	Atomic Number Element Symbol
	Unstable Nuclei

Main Groups

1 IA	2 IIA	13 IIIA	14 IVA	15 VA	16 VIA	17 VIIA	18 VIIIA	Shells
1 H							2 He	K
3 Li	4 Be	5 B	6 C	7 N	8 O	9 F	10 Ne	K–L
11 Na	12 Mg	13 Al	14 Si	15 P	16 S	17 Cl	18 Ar	K–L–M
19 K	20 Ca	31 Ga	32 Ge	33 As	34 Se	35 Br	36 Kr	–L–M–N
37 Rb	38 Sr	49 In	50 Sn	51 Sb	52 Te	53 I	54 Xe	–M–N–O
55 Cs	56 Ba	81 Tl	82 Pb	83 Bi	84 Po	85 At	86 Rn	–N–O–P
87 Fr	88 Ra							–O–P–Q

Subgroups

3 IIIB	4 IVB	5 VB	6 VIB	7 VIIB	8 VIII (1)	9 VIII (2)	10 VIII (3)	11 IB	12 IIB	Shells
21 Sc	22 Ti	23 V	24 Cr	25 Mn	26 Fe	27 Co	28 Ni	29 Cu	30 Zn	–L–M–N
39 Y	40 Zr	41 Nb	42 Mo	43 Tc	44 Ru	45 Rh	46 Pd	47 Ag	48 Cd	–M–N–O
57 La	72 Hf	73 Ta	74 W	75 Re	76 Os	77 Ir	78 Pt	79 Au	80 Hg	–N–O–P
89 Ac	104 Rf	105 Db	106 Sg	107 Bh	108 Hs	109 Mt	110 Ds	111 Rg	112	–O–P–Q

Lanthanides (Shells –N–O–P)

58	59	60	61	62	63	64	65	66	67	68	69	70	71
Ce	Pr	Nd	Pm	Sm	Eu	Gd	Tb	Dy	Ho	Er	Tm	Yb	Lu

Actinides (Shells –O–P–Q)

90	91	92	93	94	95	96	97	98	99	100	101	102	103
Th	Pa	U	Np	Pu	Am	Cm	Bk	Cf	Es	Fm	Md	No	Lr

Most Frequently Used Fundamental Constants

CODATA Recommended Values of Fundamental Constants

Quantity	Symbol and relation	Numerical value	Unit	Relative standard uncertainty
Speed of light in vacuum	c	299 792 458	m/s	Fixed by definition
Magnetic constant	$\mu_0 = 4\pi \times 10^{-7}$	$12.566370614\ldots \times 10^{-7}$	N/A^2	Fixed by definition
Electric constant	$\varepsilon_0 = 1/(\mu_0 c^2)$	$8.854187817\ldots \times 10^{-12}$	F/m	Fixed by definition
Newtonian constant of gravitation	G	$6.6742(10) \times 10^{-11}$	$m^3/(kg\,s^2)$	1.5×10^{-4}
Planck constant	h	$4.13566743(35) \times 10^{-15}$	eV s	8.5×10^{-8}
Reduced Planck constant	$\hbar = h/2\pi$	$6.58211915(56) \times 10^{-16}$	eV s	8.5×10^{-8}
Elementary charge	e	$1.60217653(14) \times 10^{-19}$	C	8.5×10^{-8}
Fine-structure constant	$\alpha = (1/4\pi\varepsilon_0)(e^2/\hbar c)$	$7.297352568(24) \times 10^{-3}$		3.3×10^{-9}
Magnetic flux quantum	$\Phi_0 = h/2e$	$2.06783372(18) \times 10^{-15}$	Wb	8.5×10^{-8}
Conductance quantum	$G_0 = 2e^2/h$	$7.748091733(26) \times 10^{-5}$	S	3.3×10^{-9}
Rydberg constant	$R_\infty = \alpha^2 m_e c/2h$	10 973 731.568525(73)	1/m	6.6×10^{-12}
Electron mass	m_e	$9.1093826(16) \times 10^{-31}$	kg	1.7×10^{-7}
Proton mass	m_p	$1.67262171(29) \times 10^{-27}$	kg	1.7×10^{-7}
Proton–electron mass ratio	m_p/m_e	1836.15267261(85)		4.6×10^{-10}
Avogadro number	N_A, L	$6.0221415(10) \times 10^{23}$		1.7×10^{-7}
Faraday constant	$F = N_A e$	96 485.3383(83)	C	8.6×10^{-8}
Molar gas constant	R	8.314472(15)	J/K	1.7×10^{-6}
Boltzmann constant	$k = R/N_A$	$1.3806505(24) \times 10^{-23}$ $8.617343(15) \times 10^{-5}$	J/K eV/K	1.8×10^{-6} 1.8×10^{-6}
Josephson constant	$K_J = 2e/h$	$483\,597.879(41) \times 10^9$	Hz/V	8.5×10^{-8}
von Klitzing constant	$R_K = h/e^2 = \mu_0 c/2\alpha$	25 812.807449(86)	Ω	3.3×10^{-9}
Bohr magneton	$\mu_B = e\hbar/2m_e$	$927.400949(80) \times 10^{-26}$ $5.788381804(39) \times 10^{-5}$	J/T eV/T	8.6×10^{-8} 6.7×10^{-9}
Atomic mass constant	$u = (1/12)m(^{12}C)$ $= (1/N_A) \times 10^{-3}$ kg	$1.66053886(28) \times 10^{-27}$	kg	1.7×10^{-7}
Bohr radius	$a_0 = \alpha/4\pi R_\infty$ $= 4\pi\varepsilon_0 \hbar^2/m_e e^2$	$0.5291772108(18) \times 10^{-10}$	m	3.3×10^{-9}
Quantum of circulation	$h/2m_e$	$3.636947550(24) \times 10^{-4}$	m^2/s	6.7×10^{-9}